PILOT ANALYSIS OF GLOBAL ECOSYSTEMS

Freshwater Systems

Carmen Revenga

Jake Brunner

Norbert Henninger

Ken Kassem

Richard Payne

CAROLLYNE HUTTER

PUBLICATIONS DIRECTOR

HYACINTH BILLINGS

PRODUCTION MANAGER

MAGGIE POWELL AND KATHY DOUCETTE

COVER DESIGN AND LAYOUT

DAVID HOSANSKY

EDITING

Each World Resources Institute Report represents a timely, scholarly treatment of a subject of public concern. WRI takes responsibility for choosing the study topics and guaranteeing its authors and researchers freedom of inquiry. It also solicits and responds to the guidance of advisory panels and expert reviewers. Unless otherwise stated, however, all the interpretation and findings set forth in WRI publications are those of the authors.

ISBN: 1-56973-460-7

Library of Congress Catalog Card No. 00-109503

Printed in the United States of America on chlorine-free paper with recycled content of 50%, 20% of which is post-consumer.

Photo Credits: Cover: Little Abiqua Creek, Oregon, Dennis A. Wentz, **Smaller ecosystem photos:** *Forests:* Digital Vision, Ltd., *Agriculture:* Philippe Berry, *Grasslands:* PhotoDisc, *Coastal:* Digital Vision, Ltd., **Prologue:** Iguaçu Falls, Argentina/Brazil, Carmen Revenga, **Human Modification of Freshwater Systems:** Dalles Dam, Oregon, Corbis Images, **Water Quantity:** Digital Vision Ltd., **Water Quality:** Florida Everglades, South Florida Water Management District, **Food — Inland Fisheries:** Nile River, Gene Wampler, **Biodiversity:** Florida Everglades, South Florida Water Management District.

Pilot Analysis of Global Ecosystems

Freshwater Systems

CARMEN REVENGA

JAKE BRUNNER

NORBERT HENNINGER

KEN KASSEM

RICHARD PAYNE

With analytical contributions from:

C. Nilsson, M. Svedmark, P. Hansson, S. Xiong, and K. Berggren, Landscape Ecology, Umeå University, Sweden (river fragmentation analysis)

Charles J. Vörösmarty and Balázs Fekete, Complex Systems Research Center, University of New Hampshire, and Wolfgang Grabs, Global Runoff Data Centre, Koblenz, Germany (global runoff analysis)

Kirsten M. J. Thompson, World Resources Institute (population projection and water scarcity analysis)

Published by World Resources Institute
Washington, DC
This report is also available at http://www.wri.org/wr2000

Pilot Analysis of Global Ecosystems (PAGE)

Project Management
Norbert Henninger, WRI
Walt Reid, WRI
Dan Tunstall, WRI
Valerie Thompson, WRI
Arwen Gloege, WRI
Elsie Velez-Whited, WRI

Agroecosystems
Stanley Wood, International Food Policy Research Institute
Kate Sebastian, International Food Policy Research Institute
Sara J. Scherr, University of Maryland

Coastal Ecosystems
Lauretta Burke, WRI
Yumiko Kura, WRI
Ken Kassem, WRI
Mark Spalding, UNEP-WCMC
Carmen Revenga, WRI
Don McAllister, Ocean Voice International

Forest Ecosystems
Emily Matthews, WRI
Richard Payne, WRI
Mark Rohweder, WRI
Siobhan Murray, WRI

Freshwater Systems
Carmen Revenga, WRI
Jake Brunner, WRI
Norbert Henninger, WRI
Ken Kassem, WRI
Richard Payne, WRI

Grassland Ecosystems
Robin White, WRI
Siobhan Murray, WRI
Mark Rohweder, WRI

A series of five technical reports, available in print and on-line at http://www.wri.org/wr2000.

AGROECOSYSTEMS

Stanley Wood, Kate Sebastian, and Sara J. Scherr, *Pilot Analysis of Global Ecosystems: Agroecosystems, A joint study by International Food Policy Research Institute and World Resources Institute*, International Food Policy Research Institute and World Resources Institute, Washington D.C.
October 2000 / paperback / ISBN 1-56973-457-7 / US$20.00

COASTAL ECOSYSTEMS

Lauretta Burke, Yumiko Kura, Ken Kassem, Mark Spalding, Carmen Revenga, and Don McAllister, *Pilot Analysis of Global Ecosystems: Coastal Ecosystems*, World Resources Institute, Washington D.C.
November 2000 / paperback / ISBN 1-56973-458-5 / US$20.00

FOREST ECOSYSTEMS

Emily Matthews, Richard Payne, Mark Rohweder, and Siobhan Murray, *Pilot Analysis of Global Ecosystems: Forest Ecosystems*, World Resources Institute, Washington D.C.
October 2000 / paperback / ISBN 1-56973-459-3 / US$20.00

FRESHWATER SYSTEMS

Carmen Revenga, Jake Brunner, Norbert Henninger, Ken Kassem, and Richard Payne *Pilot Analysis of Global Ecosystems: Freshwater Systems*, World Resources Institute, Washington D.C.
October 2000 / paperback / ISBN 1-56973-460-7 / US$20.00

GRASSLAND ECOSYSTEMS

Robin White, Siobhan Murray, and Mark Rohweder, *Pilot Analysis of Global Ecosystems: Grassland Ecosystems*, World Resources Institute, Washington D.C.
November 2000 / paperback / ISBN 1-56973-461-5 / US$20.00

The full text of each report will be available on-line at the time of publication. Printed copies may be ordered by mail from WRI Publications, P.O. Box 4852, Hampden Station, Baltimore, MD 21211, USA. To order by phone, call 1-800-822-0504 (within the United States) or 410-516-6963 or by fax 410-516-6998. Orders may also be placed on-line at http://www.wristore.com.

The agroecosystem report is also available at http://www.ifpri.org. Printed copies may be ordered by mail from the International Food Policy Research Institute, Communications Service, 2033 K Street, NW, Washington, D.C. 20006-5670, USA.

Contents

FOREWORD vii

ACKNOWLEDGMENTS ix

INTRODUCTION TO THE PILOT ANALYSIS OF GLOBAL ECOSYSTEMS .... Introduction / 1

FRESHWATER SYSTEMS: EXECUTIVE SUMMARY 1

- Scope of the Assessment
- Key Findings and Information Issues
- Conclusions

PROLOGUE: FRESHWATER SYSTEMS—WHAT THEY ARE, WHY THEY MATTER 11

HUMAN MODIFICATION OF FRESHWATER SYSTEMS 15

- Modification Of Rivers
- Changes In Groundwater Resources
- Wetland Extent and Change
- Watershed Modification
- Modification Of Freshwater Systems Information Status and Needs

WATER QUANTITY 25

- Overview
- Condition Indicators of Water Quantity
- Capacity of Freshwater Systems to Provide Water
- Water Quantity Information Status and Needs

WATER QUALITY 31

- Overview
- Condition Indicators of the Quality of Surface Waters
- Condition Indicators of Groundwater Quality
- Capacity of Freshwater Systems to Provide Clean Water
- Water Quality Information Status and Needs

FOOD—INLAND FISHERIES 41

- Status and Trends in Inland Fisheries
- Pressures on Inland Fishery Resources
- Condition Indicators of Fish Production
- Capacity of Freshwater System to Provide Food
- Inland Fisheries Information Status and Needs

BIODIVERSITY 49

Overview
Indicators of Biological Value
Condition Indicators of Biodiversity
Capacity of Freshwater Systems to Sustain Biodiversity
Biodiversity Information Status and Needs

TABLES

Table 1. Alteration of Freshwater Systems Worldwide 16
Table 2. Large Dams and Storage Capacity of Large Reservoirs by Continent 17
Table 3. Global Annual Renewable Water Supply Per Person in 1995 and Projections for 2025 27
Table 4. Trends in U.S. Stream Water Quality, 1980–89 35
Table 5. Changes in Fish Species Composition and Fisheries for Selected Rivers, Lakes, and Inland Seas 45
Table 6. Species Richness by Ecosystem 50
Table 7. Watersheds with High Fish Species Richness and Endemism 51

FIGURES

Figure 1. Trading Biodiversity for Export Earnings: The Changing Lake Victoria Fishery (Kenya Only) 12
Figure 2. Skjern River Floodplain: Marshland and Meadow Area in 1871 and 1987 13
Figure 3. Statistical distribution of BOD by Continent, 1976–90 34
Figure 4. Inland Capture Fisheries by Continent, 1984–97 42
Figure 5. Top Ten Producing Countries, 1997 43
Figure 6. Global Inland Capture and Freshwater Aquaculture Growth, 1984–97 44
Figure 7. Commercial Landings of Salmon and Steelhead in the Columbia River, 1886–1998 46
Figure 8. Globally Outstanding Freshwater Ecoregions in North America 52

BOXES

Box 1. What Is at Stake and What Are the Trade-offs: Okavango Delta 20
Box 2. Index of Watershed Indicators from the U.S. Environmental Protection Agency 32
Box 3. Threats and Issues Facing Inland Fisheries by Continent 47

MAPS

Map 1. River Channel Fragmentation and Flow Regulation
Map 2. Number of Large Dams under Construction by River Basin as of 1998
Map 3. Residence Time of Continental Runoff by River Basin
Map 4. Africa: Wetlands, Dams, and Ramsar Sites
Maps 5a and 5b. United States: Historical Wetland Loss by State, 1780s–1980s and United States: Net Change in Wetland Area, 1982–92
Map 6. Percentage of Cropland Area by River Basin
Map 7. Percentage of Urban and Industrial Land Use by River Basin
Maps 8a and 8b. Europe and the Middle East: Intensive Agricultural Land Use by Subbasin and Insular Southeast Asia: Intensive Agricultural Land Use by Subbasin
Map 9. Annual Renewable Water Supply Per Person by River Basin, 1995
Map 10. Projected Annual Renewable Water Supply Per Person by River Basin, 2025
Map 11. Annual Renewable Water Supply and Dry Season Flow by River Basin
Map 12. Trends in Inland Capture Fisheries by Country, 1984–97
Map 13. Important Areas and Ecoregions for Freshwater Biodiversity
Map 14. Fish Species Richness and Endemism by River Basin
Map 15. Amphibian Census Sites and Decline Index
Map 16. Imperiled Freshwater Fish and Herpetofauna in North American Freshwater Ecoregions
Map 17. Number of Species Introductions into Inland Waters by Country
Map 18. Zebra Mussel Expansion and Water Hyacinth Presence in the United States

BIBLIOGRAPHY 59

Foreword

Earth's ecosystems and its peoples are bound together in a grand and complex symbiosis. We depend on ecosystems to sustain us, but the continued health of ecosystems depends, in turn, on our use and care. Ecosystems are the productive engines of the planet, providing us with everything from the water we drink to the food we eat and the fiber we use for clothing, paper, or lumber. Yet, nearly every measure we use to assess the health of ecosystems tells us we are drawing on them more than ever and degrading them, in some cases at an accelerating pace.

Our knowledge of ecosystems has increased dramatically in recent decades, but it has not kept pace with our ability to alter them. Economic development and human well-being will depend in large part on our ability to manage ecosystems more sustainably. We must learn to evaluate our decisions on land and resource use in terms of how they affect the capacity of ecosystems to sustain life — not only human life, but also the health and productive potential of plants, animals, and natural systems.

A critical step in improving the way we manage the earth's ecosystems is to take stock of their extent, their condition, and their capacity to provide the goods and services we will need in years to come. To date, no such comprehensive assessment of the state of the world's ecosystems has been undertaken.

The Pilot Analysis of Global Ecosystems (PAGE) begins to address this gap. This study is the result of a remarkable collaborative effort between the World Resources Institute (WRI), the International Food Policy Research Institute (IFPRI), intergovernmental organizations, agencies, research institutes, and individual experts in more than 25 countries worldwide. The PAGE compares information already available on a global scale about the condition of five major classes of ecosystems: agroecosystems, coastal areas, forests, freshwater systems, and grasslands. IFPRI led the agroecosystem analysis, while the others were led by WRI. The pilot analysis examines not only the quantity and quality of outputs but also the biological basis for production, including soil and water condition, biodiversity, and changes in land use over time. Rather than looking just at marketed products, such as food and timber, the study also analyses the condition of a broad array of ecosystem goods and services that people need, or enjoy, but do not buy in the marketplace.

The five PAGE reports show that human action has profoundly changed the extent, condition, and capacity of all major ecosystem types. Agriculture has expanded at the expense of grasslands and forests, engineering projects have altered the hydrological regime of most of the world's major rivers, settlement and other forms of development have converted habitats around the world's coastlines. Human activities have adversely altered the earth's most important biogeochemical cycles — the water, carbon, and nitrogen cycles — on which all life forms depend. Intensive management regimes and infrastructure development have contributed positively to providing some goods and services, such as food and fiber from forest plantations. They have also led to habitat fragmentation, pollution, and increased ecosystem vulnerability to pest attack, fires, and invasion by nonnative species. Information is often incomplete and the picture confused, but there are many signs that the overall capacity of ecosystems to continue to produce many of the goods and services on which we depend is declining.

The results of the PAGE are summarized in *World Resources 2000–2001*, a biennial report on the global environment published by the World Resources Institute in partnership with the United Nations Development Programme, the United Nations Environment Programme, and the World Bank. These institutions have affirmed their commitment to making the viability of the world's ecosystems a critical development priority for the 21st century. WRI and its partners began work with a conviction that the challenge of managing earth's ecosystems — and the consequences of failure — will increase significantly in coming decades. We end with a keen awareness that the scientific knowledge and political will required to meet this challenge are often lacking today. To make sound ecosystem management decisions in the future, significant changes are needed in the way we use the knowledge and experience at hand, as well as the range of information brought to bear on resource management decisions.

A truly comprehensive and integrated assessment of global ecosystems that goes well beyond our pilot analysis is necessary to meet information needs and to catalyze regional and local assessments. Planning for such a Millennium Ecosystem Assessment is already under way. In 1998, representatives from international scientific and political bodies began to explore the merits of, and recommend the structure for, such an assessment. After consulting for a year and considering the preliminary findings of the PAGE report, they concluded that an international scientific assessment of the present and likely future condition of the world's ecosystems was both feasible and urgently needed. They urged local, national, and international institutions to support the effort as stakeholders, users, and sources of expertise. If concluded successfully, the Millennium Ecosystem Assessment will generate new information, integrate current knowledge, develop methodological tools, and increase public understanding.

Human dominance of the earth's productive systems gives us enormous responsibilities, but great opportunities as well. The challenge for the 21st century is to understand the vulnerabilities and resilience of ecosystems, so that we can find ways to reconcile the demands of human development with the tolerances of nature.

We deeply appreciate support for this project from the Australian Centre for International Agricultural Research, The David and Lucile Packard Foundation, The Netherlands Ministry of Foreign Affairs, the Swedish International Development Cooperation Agency, the United Nations Development Programme, the United Nations Environment Programme, the Global Bureau of the United States Agency for International Development, and The World Bank.

A special thank you goes to the AVINA Foundation, the Global Environment Facility, and the United Nations Fund for International Partnerships for their early support of PAGE and the Millennium Ecosystem Assessment, which was instrumental in launching our efforts.

Jonathan Lash
President
World Resources Institute

Acknowledgments

The World Resources Institute and the International Food Policy Research Institute would like to acknowledge the members of the Millennium Assessment Steering Committee, who generously gave their time, insights, and expert review comments in support of the Pilot Analysis of Global Ecosystems.

Edward Ayensu, Ghana; Mark Collins, United Nations Environment Programme-World Conservation Monitoring Centre (UNEP -WCMC), United Kingdom; Angela Cropper, Trinidad and Tobago; Andrew Dearing, World Business Council for Sustainable Development (WBCSD); Janos Pasztor, UNFCCC; Louise Fresco, FAO; Madhav Gadgil, Indian Institute of Science, Bangalore, India; Habiba Gitay, Australian National University, Australia; Gisbert Glaser, UNESCO; Zuzana Guziova, Ministry of the Environment, Slovak Republic; He Changchui, FAO; Calestous Juma, Harvard University; John Krebs, National Environment Research Council, United Kingdom; Jonathan Lash, World Resources Institute; Roberto Lenton, UNDP; Jane Lubchenco, Oregon State University; Jeffrey McNeely, World Conservation Union (IUCN), Switzerland; Harold Mooney, International Council for Science (ICSU); Ndegwa Ndiangui, Convention to Combat Desertification; Prabhu L. Pingali, CIMMYT; Per Pinstrup-Andersen, Consultative Group on International Agricultural Research; Mario Ramos, Global Environment Facility; Peter Raven, Missouri Botanical Garden; Walter Reid, Secretariat; Cristian Samper, Instituto Alexander Von Humboldt, Colombia; José Sarukhán, CONABIO, Mexico; Peter Schei, Directorate for Nature Management, Norway; Klaus Töpfer, UNEP; José Galízia Tundisi, International Institute of Ecology, Brazil; Robert Watson, World Bank; Xu Guanhua, Ministry of Science and Technology, People's Republic of China; A.H. Zakri, Universiti Kebangsaan Malaysia, Malaysia.

The Pilot Analysis of Global Ecosystems would not have been possible without the data provided by numerous institutions and agencies. The authors of the freshwater systems analysis wish to express their gratitude for the generous cooperation and invaluable information they received from the following organizations:

BirdLife International; Center for Environmental Systems Research, University of Kassel, Germany; Center for International Earth Science Information Network (CIESIN); Complex Systems Research Center, University of New Hampshire; Convention on Wetlands; COWI Consulting Engineers and Planners AS, Denmark; Declining Amphibian Populations Task Force; DHI Water & Environment, Denmark; European Environment Agency; Fisheries Department, Food and Agriculture Organization of the United Nations (FAO); Global Runoff Data Centre, Koblenz, Germany; Land and Water Development Division, FAO; Landscape Ecology, Umeå University, Sweden; U.S. National Oceanic and Atmospheric Administration – National Geophysical Data Center (NOAA-NGDC); National Resources Conservation Service, U.S. Department of Agriculture (USDA); National Wetlands Inventory, United States Fish and Wildlife Service; Oregon Department of Fish and Wildlife, U.S.A; State of Ohio Environmental Protection Agency, U.S.A.; United States Geological Survey (USGS); Washington Department of Fish and Wildlife, U.S.A.; World Conservation Monitoring Centre (WCMC); World Wildlife Fund-U.S. (WWF-U.S.)

The authors would also like to express their gratitude to the many individuals who contributed information and advice, attended expert workshops, and reviewed successive drafts of this report.

Robin Abell, WWF-U.S.; Devin Bartley, Fisheries Department, FAO; Amy Benson, USGS; Ger Bergkamp, IUCN- The World Conservation Union; Stephen J. Brady, Natural Resources Conservation Service, USDA; Jesslyn Brown, USGS/EROS Data Center; Morley Brownstein, Health Canada; Cynthia Carey, Department of Biology, University of Colorado; John Cooper, Environment Canada; Thomas E. Dahl, National Wetlands Inventory, U.S. Fish and Wildlife Service; Nick Davidson, Convention on Wetlands; Jean-Marc Faurès, Land and Water Development Division, FAO; Stephen Foster, British Geological Survey; Andy Fraser, Environment Canada; Scott Frazier, Wetlands International; Brij Gopal, Jawaharlal Nehru University, New Delhi, India; Jippe Hoogeveen, Land and Water Development Division, FAO; Colette Jacono, USGS; Jim Kapetsky, Fisheries Department, FAO; James Karr, University of Washington; Les Kaufman, Marine Program, Boston University; Kim Martz, USGS; Don

McAllister, Ocean Voice International; Timothy L. Miller, USGS; Peter Moyle, Wildlife, Fish, and Conservation Biology, University of California, Davis; Tom Neill, Oregon Department of Fish and Wildlife; Christer Nilsson, Umeå University, Sweden; Kim W. Olesen, DHI Water & Environment, Denmark; Francisco Olivera, Center for Research in Water Resources, University of Texas at Austin; Chales R. O'Neill, National Zebra Mussel Information Clearinghouse, New York Sea Grant; Sandra Postel, Global Water Policy Project; Edward T. Rankin, State of Ohio Environmental Protection Agency; Corinna Ravilious, WCMC; Ilze Reiss, Environment Canada; Hans H. Riber, COWI Consulting Engineers and Planners AS, Denmark; Steve Rothert, International Rivers Network; Robert Rusin, Goddard Space Flight Center, NASA; Dork Sahagian, IGBP/GAIM, University of New Hampshire; John R. Sauer, USGS; Teresa Scott, Washington Department of Fish and Wildlife; Igor Shiklomanov, State Hydrological Institute, St. Petersburg, Russia; Robert Slater, Environment Canada; Charles Spooner, U.S. Environmental Protection Agency; Bruce Stein, The Nature Conservancy; Melanie J. Stiassny, American Museum of Natural History; Greg Thompson, Environment Canada, Niels Thyssen, European Environment Agency; Joshua Viers, Dept. of Environmental Science and Policy, University of California, Davis; Zipangani M. Vokhiwa, Ministry of Research and Environmental Affairs, Malawi; Charles Vörösmarty, University of New Hampshire; David Wilcove, Environmental Defense.

We also wish to thank the many individuals at WRI who were generous with their help as this report progressed: Tony Janetos, Yumiko Kura, Gregory Mock, and Dan Tunstall. Kirsten Thompson and Johnathan Kool worked tirelessly in the GIS lab to produce the PAGE maps. Hyacinth Billings, Kathy Doucette, Carollyne Hutter, and Maggie Powell guided the report through production with their usual calm skill, and provided editorial and design assistance.

Introduction to the Pilot Analysis of Global Ecosystems

PEOPLE AND ECOSYSTEMS

The world's economies are based on the goods and services derived from ecosystems. Human life itself depends on the continuing capacity of biological processes to provide their multitude of benefits. Yet, for too long in both rich and poor countries, development priorities have focused on how much humanity can take from ecosystems, and too little attention has been paid to the impact of our actions. We are now experiencing the effects of ecosystem decline in numerous ways: water shortages in the Punjab, India; soil erosion in Tuva, Russia; fish kills off the coast of North Carolina in the United States; landslides on the deforested slopes of Honduras; fires in the forests of Borneo and Sumatra in Indonesia. The poor, who often depend directly on ecosystems for their livelihoods, suffer most when ecosystems are degraded.

A critical step in managing our ecosystems is to take stock of their extent, their condition, and their capacity to continue to provide what we need. Although the information available today is more comprehensive than at any time previously, it does not provide a complete picture of the state of the world's ecosystems and falls far short of management and policy needs. Information is being collected in abundance but efforts are often poorly coordinated. Scales are noncomparable, baseline data are lacking, time series are incomplete, differing measures defy integration, and different information sources may not know of each other's relevant findings.

OBJECTIVES

The Pilot Analysis of Global Ecosystems (PAGE) is the first attempt to synthesize information from national, regional, and global assessments. Information sources include state of the environment reports; sectoral assessments of agriculture, forestry, biodiversity, water, and fisheries, as well as national and global assessments of ecosystem extent and change; scientific research articles; and various national and international data sets. The study reports on five major categories of ecosystems:

- Agroecosystems;
- Coastal ecosystems;
- Forest ecosystems;
- Freshwater systems;
- Grassland ecosystems.

These ecosystems account for about 90 percent of the earth's land surface, excluding Greenland and Antarctica. PAGE results are being published as a series of five technical reports, each covering one ecosystem. Electronic versions of the reports are posted on the Website of the World Resources Institute [http://www.wri.org/wr2000] and the agroecosystems report also is available on the Website of the International Food Policy Research Institute [http://www/ifpri.org].

The primary objective of the pilot analysis is to provide an overview of ecosystem condition at the global and continental levels. The analysis documents the extent and distribution of the five major ecosystem types and identifies ecosystem change over time. It analyzes the quantity and quality of ecosystem goods and services and, where data exist, reviews trends relevant to the production of these goods and services over the past 30 to 40 years. Finally, PAGE attempts to assess the capacity of ecosystems to continue to provide goods and services, using measures of biological productivity, including soil and water conditions, biodiversity, and land use. Wherever possible, information is presented in the form of indicators and maps.

A second objective of PAGE is to identify the most serious information gaps that limit our current understanding of ecosystem condition. The information base necessary to assess ecosystem condition and productive capacity has not improved in recent years, and may even be shrinking as funding for environmental monitoring and record-keeping diminishes in some regions.

Most importantly, PAGE supports the launch of a Millennium Ecosystem Assessment, a more ambitious, detailed, and integrated assessment of global ecosystems that will provide a firmer basis for policy- and decision-making at the national and subnational scale.

AN INTEGRATED APPROACH TO ASSESSING ECOSYSTEM GOODS AND SERVICES

Ecosystems provide humans with a wealth of goods and services, including food, building and clothing materials, medicines, climate regulation, water purification, nutrient cycling, recreation opportunities, and amenity value. At present, we tend to manage ecosystems for one dominant good or service, such as grain, fish, timber, or hydropower, without fully realizing the trade-offs we are making. In so doing, we may be sacrificing goods or services more valuable than those we receive — often those goods and services that are not yet valued in the market, such as biodiversity and flood control. An integrated ecosystem approach considers the entire range of possible goods and services a given ecosystem provides and attempts to optimize the benefits that society can derive from that ecosystem and across ecosystems. Its purpose is to help make trade-offs efficient, transparent, and sustainable.

Such an approach, however, presents significant methodological challenges. Unlike a living organism, which might be either healthy or unhealthy but cannot be both simultaneously, ecosystems can be in good condition for producing certain goods and services but in poor condition for others. PAGE attempts to evaluate the condition of ecosystems by assessing separately their capacity to provide a variety of goods and services and examining the trade-offs humans have made among those goods and services. As one example, analysis of a particular region might reveal that food production is high but, because of irrigation and heavy fertilizer application, the ability of the system to provide clean water has been diminished.

Given data inadequacies, this systematic approach was not always feasible. For each of the five ecosystems, PAGE researchers, therefore, focus on documenting the extent and distribution of ecosystems and changes over time. We develop indicators of ecosystem condition — indicators that inform us about the current provision of goods and services and the likely capacity of the ecosystem to continue providing those goods and services. Goods and services are selected on the basis of their perceived importance to human development. Most of the ecosystem studies examine food production, water quality and quantity, biodiversity, and carbon sequestration. The analysis of forests also studies timber and woodfuel production; coastal and grassland studies examine recreational and tourism services; and the agroecosystem study reviews the soil resource as an indicator of both agricultural potential and its current condition.

PARTNERS AND THE RESEARCH PROCESS

The Pilot Analysis of Global Ecosystems was a truly international collaborative effort. The World Resources Institute and the International Food Policy Research Institute carried out their research in partnership with numerous institutions worldwide (see *Acknowledgments*). In addition to these partnerships, PAGE researchers relied on a network of international experts for ideas, comments, and formal reviews. The research process included meetings in Washington, D.C., attended by more than 50 experts from developed and developing countries. The meetings proved invaluable in developing the conceptual approach and guiding the research program toward the most promising indicators given time, budget, and data constraints. Drafts of PAGE reports were sent to over 70 experts worldwide, presented and critiqued at a technical meeting of the Convention on Biological Diversity in Montreal (June, 1999) and discussed at a Millennium Assessment planning meeting in Kuala Lumpur, Malaysia (September, 1999). Draft PAGE materials and indicators were also presented and discussed at a Millennium Assessment planning meeting in Winnipeg, Canada, (September, 1999) and at the meeting of the Parties to the Convention to Combat Desertification, held in Recife, Brazil (November, 1999).

KEY FINDINGS

Key findings of PAGE relate both to ecosystem condition and the information base that supported our conclusions.

The Current State of Ecosystems

The PAGE reports show that human action has profoundly changed the extent, distribution, and condition of all major ecosystem types. Agriculture has expanded at the expense of grasslands and forests, engineering projects have altered the hydrological regime of most of the world's major rivers, settlement and other forms of development have converted habitats around the world's coastlines.

The picture we get from PAGE results is complex. Ecosystems are in good condition for producing some goods and services but in poor condition for producing others. Overall, however, there are many signs that the capacity of ecosystems to continue to produce many of the goods and services on which we depend is declining. Human activities have significantly disturbed the global water, carbon, and nitrogen cycles on which all life depends. Agriculture, industry, and the spread of human settlements have permanently converted extensive areas of natural habitat and contributed to ecosystem degradation through fragmentation, pollution, and increased incidence of pest attacks, fires, and invasion by nonnative species.

The following paragraphs look across ecosystems to summarize trends in production of the most important goods and services and the outlook for ecosystem productivity in the future.

Food Production

Food production has more than kept pace with global population growth. On average, food supplies are 24 percent higher per person than in 1961 and real prices are 40 percent lower. Production is likely to continue to rise as demand increases in the short to medium term. Long-term productivity, however, is threatened by increasing water scarcity and soil degradation, which is now severe enough to reduce yields on about 16 percent of agricultural land, especially cropland in Africa and Central America and pastures in Africa. Irrigated agriculture, an important component in the productivity gains of the Green Revolution, has contributed to waterlogging and salinization, as well as to the depletion and chemical contamination of surface and groundwater supplies. Widespread use of pesticides on crops has lead to the emergence of many pesticide-resistant pests and pathogens, and intensive livestock production has created problems of manure disposal and water pollution. Food production from marine fisheries has risen sixfold since 1950 but the rate of increase has slowed dramatically as fisheries have been overexploited. More than 70 percent of the world's fishery resources for which there is information are now fully fished or overfished (yields are static or declining). Coastal fisheries are under threat from pollution, development, and degradation of coral reef and mangrove habitats. Future increases in production are expected to come largely from aquaculture.

Water Quantity

Dams, diversions, and other engineering works have transformed the quantity and location of freshwater available for human use and sustaining aquatic ecosystems. Water engineering has profoundly improved living standards, by providing fresh drinking water, water for irrigation, energy, transport, and flood control. In the twentieth century, water withdrawals have risen at more than double the rate of population increase and surface and groundwater sources in many parts of Asia, North Africa, and North America are being depleted. About 70 percent of water is used in irrigation systems where efficiency is often so low that, on average, less than half the water withdrawn reaches crops. On almost every continent, river modification has affected the flow of rivers to the point where some no longer reach the ocean during the dry season. Freshwater wetlands, which store water, reduce flooding, and provide specialized biodiversity habitat, have been reduced by as much as 50 percent worldwide. Currently, almost 40 percent of the world's population experience serious water shortages. Water scarcity is expected to grow dramatically in some regions as competition for water grows between agricultural, urban, and commercial sectors.

Water Quality

Surface water quality has improved with respect to some pollutants in developed countries but water quality in developing countries, especially near urban and industrial areas, has worsened. Water is degraded directly by chemical or nutrient pollution, and indirectly when land use change increases soil erosion or reduces the capacity of ecosystems to filter water. Nutrient runoff from agriculture is a serious problem around the world, resulting in eutrophication and human health hazards in coastal regions, especially in the Mediterranean, Black Sea, and northwestern Gulf of Mexico. Water-borne diseases caused by fecal contamination of water by untreated sewage are a major source of morbidity and mortality in the developing world. Pollution and the introduction of non-native species to freshwater ecosystems have contributed to serious declines in freshwater biodiversity.

Carbon Storage

The world's plants and soil organisms absorb carbon dioxide (CO_2) during photosynthesis and store it in their tissues, which helps to slow the accumulation of CO_2 in the atmosphere and mitigate climate change. Land use change that has increased production of food and other commodities has reduced the net capacity of ecosystems to sequester and store carbon. Carbon-rich grasslands and forests in the temperate zone have been extensively converted to cropland and pasture, which store less carbon per unit area of land. Deforestation is itself a significant source of carbon emissions, because carbon stored in plant tissue is released by burning and accelerated decomposition. Forests currently store about 40 percent of all the carbon held in terrestrial ecosystems. Forests in the northern hemisphere are slowly increasing their storage capacity as they regrow after historic clearance. This gain, however, is more than offset by deforestation in the tropics. Land use change accounts for about 20 percent of anthropogenic carbon emissions to the atmosphere. Globally, forests today are a net source of carbon.

Biodiversity

Biodiversity provides many direct benefits to humans: genetic material for crop and livestock breeding, chemicals for medicines, and raw materials for industry. Diversity of living organisms and the abundance of populations of many species are also critical to maintaining biological services, such as pollination and nutrient cycling. Less tangibly, but no less importantly, diversity in nature is regarded by most people as valuable in its own right, a source of aesthetic pleasure, spiritual solace, beauty, and wonder. Alarming losses in global biodiversity have occurred over the past century. Most are the result of habitat destruction. Forests, grasslands, wetlands, and mangroves have been extensively converted to other uses; only tundra, the Poles, and deep-sea ecosystems have experienced relatively little change. Biodiversity has suffered as agricultural land, which supports far less biodiversity than natural forest, has expanded primarily at the expense of forest areas. Biodiversity is also diminished by intensification, which reduces the area allotted to hedgerows, copses, or wildlife corridors and displaces traditional varieties of seeds with modern high-yield, but genetically uniform, crops. Pollution, overexploitation, and competition from invasive species represent further threats to biodiversity. Freshwater ecosystems appear to be the most severely degraded overall, with an estimated 20 percent of freshwater fish species becoming extinct, threatened, or endangered in recent decades.

Information Status and Needs

Ecosystem Extent and Land Use Characterization

Available data proved adequate to map approximate ecosystem extent for most regions and to estimate historic change in grassland and forest area by comparing current with potential vegetation cover. PAGE was able to report only on recent changes in ecosystem extent at the global level for forests and agricultural land.

PAGE provides an overview of human modifications to ecosystems through conversion, cultivation, firesetting, fragmentation by roads and dams, and trawling of continental shelves. The study develops a number of indicators that quantify the degree of human modification but more information is needed to document adequately the nature and rate of human modifications to ecosystems. Relevant data at the global level are incomplete and some existing data sets are out of date.

Perhaps the most urgent need is for better information on the spatial distribution of ecosystems and land uses. Remote sensing has greatly enhanced our knowledge of the global extent of vegetation types. Satellite data can provide invaluable information on the spatial pattern and extent of ecosystems, on their physical structure and attributes, and on rates of change in the landscape. However, while gross spatial changes in vegetation extent can be monitored using coarse-resolution satellite data, quantifying land cover change at the national or subnational level requires high-resolution data with a resolution of tens of meters rather than kilometers.

Much of the information that would allow these needs to be met, at both the national and global levels, already exists, but is not yet in the public domain. New remote sensing techniques and improved capabilities to manage complex global data sets mean that a complete satellite-based global picture of the earth could now be made available, although at significant cost. This information would need to be supplemented by extensive ground-truthing, involving additional costs. If sufficient resources were committed, fundamentally important information on ecosystem extent, land cover, and land use patterns around the world could be provided at the level of detail needed for national planning. Such information would also prove invaluable to international environmental conventions, such as those dealing with wetlands, biological diversity, desertification, and climate change, as well as the international agriculture, forest, and fishery research community.

Ecosystem Condition and Capacity to Provide Goods and Services

In contrast to information on spatial extent, data that can be used to analyze ecosystem condition are often unavailable or incomplete. Indicator development is also beset by methodological difficulties. Traditional indicators, for example, those relating to pressures on environments, environmental status, or societal responses (pressure-state-response model indicators) provide only a partial view and reveal little about the underlying capacity of the ecosystem to deliver desired goods and services. Equally, indicators of human modification tell us about changes in land use or biological parameters, but do not necessarily inform us about potentially positive or negative outcomes.

Ecosystem conditions tend to be highly site-specific. Information on rates of soil erosion or species diversity in one area may have little relevance to an apparently similar system a few miles away. It is expensive and challenging to monitor and synthesize site-specific data and present it in a form suitable for national policy and resource management decisions. Finally, even where data are available, scientific understanding of how changes in biological systems will affect goods and services is limited. For example, experimental evidence shows that loss of biological diversity tends to reduce the resilience of a system to perturbations, such as storms, pest outbreaks, or climate change. But scientists are not yet able to quantify how much resilience is lost as a result of the loss of biodiversity in a particular site or how that loss of resilience might affect the long-term production of goods and services.

Overall, the availability and quality of information tend to match the recognition accorded to various goods and services by markets. Generally good data are available for traded goods, such as grains, fish, meat, and timber products and some of the more basic relevant productivity factors, such as fertilizer application rates, water inputs, and yields. Data on products that are exchanged in informal markets, or consumed directly, are patchy and often modeled. Examples include fish landings from artisanal fisheries, woodfuels, subsistence food crops and livestock, and nonwood forest products. Information on the biological factors that support production of these goods — including size of fish spawning stocks, biomass densities, subsistence food yields, and forest food harvests — are generally absent.

The future capacity (long-term productivity) of ecosystems is influenced by biological processes, such as soil formation, nutrient cycling, pollination, and water purification and cycling. Few of these environmental services have, as yet, been accorded economic value that is recognized in any functioning market. There is a corresponding lack of support for data collection and monitoring. This is changing in the case of carbon storage and cycling. Interest in the possibilities of carbon trading mechanisms has stimulated research and generated much improved data on carbon stores in terrestrial ecosystems and the dimensions of the global carbon cycle. Few comparable datasets exist for elements such as nitrogen or sulfur, despite their fundamental importance in maintaining living systems.

Although the economic value of genetic diversity is growing, information on biodiversity is uniformly poor. Baseline and trend data are largely lacking; only an estimated 15 to 20 percent of the world's species have been identified. The OECD Megascience Forum has launched a new international program to accelerate the identification and cataloging of species around the world. This information will need to be supplemented with improved data on species population trends and the numbers and abundance of invasive species. Developing databases on population trends (and threat status) is likely to be a major challenge, because most countries still need to establish basic monitoring programs.

The PAGE divides the world's ecosystems to examine them at a global scale and think in broad terms about the challenges of managing them sustainably. In reality, ecosystems are linked by countless flows of material and human actions. The PAGE analysis does not make a distinction between natural and managed ecosystems; human intervention affects all ecosystems to some degree. Our aim is to take a first step toward understanding the collective impacts of those interventions on the full range of goods and services that ecosystems provide. We conclude that we lack much of the baseline information necessary to determine ecosystem conditions at a global, regional or, in many instances, even a local scale. We also lack systematic approaches necessary to integrate analyses undertaken at different locations and spatial scales.

Finally, it should be noted that PAGE looks at past trends and current status, but does not try to project future situations where, for example, technological development might increase dramatically the capacity of ecosystems to deliver the goods and services we need. Such considerations were beyond the scope of the study. However, technologies tend to be developed and applied in response to market-related opportunities. A significant challenge is to find those technologies, such as integrated pest management and zero tillage cultivation practices in the case of agriculture, that can simultaneously offer market-related as well as environmental benefits. It has to be recognized, nonetheless, that this type of "win-win" solution may not always be possible. In such cases, we need to understand the nature of the trade-offs we must make when choosing among different combinations of goods and services. At present our knowledge is often insufficient to tell us where and when those trade-offs are occurring and how we might minimize their effects.

Freshwater Systems: Executive Summary

The constant cycle of water between the oceans, atmosphere, and land sustains life on Earth. All organisms on the planet need water to survive. Without water, microorganisms that decompose organic matter could not exist, interrupting the ecological loops of matter and energy and shutting down all ecosystems.

Freshwater systems are created by water that enters the terrestrial environment as precipitation and flows both above and below ground toward the sea. These systems encompass a wide range of habitats, including rivers, lakes, and wetlands, and the riparian zones associated with them. Their boundaries are constantly changing with the seasonality in the hydrological cycle. Their environmental benefits and costs are distributed widely across time and space, because of the complex interactions between climate, surface water and groundwater, and coastal marine areas.

This analysis concentrates on the terrestrial water that is most accessible to humans: the water in rivers, lakes, and wetlands. Humans also rely heavily on groundwater, which is the only source of fresh water in some parts of the world. However, this paper will not focus as extensively on groundwater, in part because the data on this resource are scarce.

Scope of the Assessment

This study analyzes quantitative and qualitative information and develops selected indicators of the condition of the world's freshwater systems. The condition is defined as the current and future capacity of the systems to continue providing the full range of goods and services needed or valued by humans.

Where available, we use global data sets to illustrate key indicators. In cases in which global data are not available, we use regional- and national-level information to illustrate important concepts, indicators, trends, and issues. Sometimes, local-level case studies have been used to illustrate trends that appear to be important but for which national or global data do not exist.

On a global scale, only limited information is available on the condition of the world's freshwater systems. Often the spatial resolution and temporal domains for different parts of the world are poorly harmonized. Most developing countries lack environmental monitoring programs for freshwater systems. Even developed countries with data-collection systems on hydrology, species, habitats, and physical and chemical parameters of water quality have done little to develop indicators that measure important ecological processes of freshwater systems, such as water purification and aquifer recharge. More also needs to be done to integrate data for entire watersheds, from water supply and consumption to land use and biodiversity.

Our analysis looks at measures that show the degree of human intervention in the hydrological cycle and what we know concerning three important goods and services provided by freshwater systems: water, food, and biodiversity. These goods and services were chosen partly on the advice of a wide range of freshwater experts and partly because of data availability. The data and indicators presented in this pilot analysis focus on the following:

- Human modification of freshwater systems. (These include all physical changes in the hydrological cycle, especially river and stream corrections, flood control by dams, conversion of wetlands, and land-use changes in the entire watershed—all of which are changing runoff characteristics).
- Water quantity (i.e., availability).
- Water quality.
- Food (fish in particular).
- Biodiversity.

We use these indicators to identify existing data, highlight characteristics of ideal indicators to measure the capacity of freshwater ecosystems, and point out data and information needs. These in turn will become useful inputs for the Millennium Ecosystem Assessment.

Even though food production from irrigated crops is intimately related to water availability, and agriculture is the biggest user of water, this report does not assess the condition of freshwater systems for agriculture production. A separate Pilot Analysis of Global Ecosystems (PAGE) report on agroecosystems assesses agriculture production in detail. Moreover, this report on freshwater systems does not cover additional important services derived from freshwater systems such as hydropower, transportation, and recreation.

Hydropower electricity production plays a significant role in the overall energy output of many countries. Of the total electricity generated in the world, hydropower accounts for 18 percent and in 18 countries, including Brazil, Norway, Burundi, and Laos, it generates 90 percent or more of the electricity (Gleick 1998:276–280).

Since historical times, transportation has been a crucial service that humans have derived from freshwater systems. Rivers have been harnessed as routes for exploring, colonizing, and settling new areas, as well as transporting goods and communicating. In western Europe, for example, inland waterways transport almost 8 percent of all inland freight (EEA 1995:441).

Tourism and recreation, as well as the more subtle spiritual and aesthetic qualities of freshwater systems, constitute perhaps the most important omission from this study. Society places a high value on freshwater recreational activities, such as boating, fishing, hunting, birdwatching, and swimming. These services generate billions of dollars in direct and indirect revenue in many developed and developing countries. Exact figures are hard to calculate because of the dispersed nature of these activities. In the United States alone, however, 30 million anglers went freshwater fishing in 1996, expending US$24.5 billion on trips and equipment (UFWS 1996:8).

The spiritual and aesthetic qualities of freshwater systems cannot readily be captured by the kind of quantitative analysis presented here. For two reasons, this study does not consider data on tourism revenues, which some analysts have used as proxy measures of human appreciation. First, the very concept of analyzing freshwater systems goods and services is essentially utilitarian, whereas emotional commitment to these systems as things of beauty or intrinsic value is essentially normative. Second, any attempt to develop quantitative indicators of such intangible issues risks removing them from their proper arena of political, moral, and cultural debate.

Finally, even though groundwater resources play a critical role in many regions of the world by providing potable, industrial, and irrigation water, global data on this resource are scarce and dispersed among national agencies. This analysis, therefore, provides a general overview of the issues affecting groundwater resources and their condition.

Key Findings and Information Issues

The following tables (pp. 3–7) summarize key findings of the study regarding the condition of freshwater systems, as well as the quality and availability of data.

Human Modification of Freshwater Systems

PAGE MEASURES AND INDICATORS	DATA SOURCES AND COMMENTS
Historical alteration of freshwater systems worldwide	Compilation of data from the following sources: based on Naiman et al. 1995 as adapted from L'vovich and White 1990. Additional data from Shiklomanov 1997, ICOLD 1998, Avakyan and Iakovleva 1998, and IJHD 1998.
River channel fragmentation and flow regulation	Analysis was done by C. Nilsson, M. Svedmark, P. Hansson, S. Xiong, and K. Berggren, Landscape Ecology, Umeå University, Sweden. Additional data and analysis from Dynesius and Nilsson 1994. Rivers assessed are those with a historical record of more than 350 virgin mean annual discharge. Based on available information on dams and other flow regulations. Not all regions of the world were assessed.
Number of large dams under construction by river basin	IJHD 1998. This data set includes only reported dams over 60 meters high that are currently under construction, aggregated by river basin.
Residence time of continental runoff by river basin	Vörösmarty et al. 1997a. This indicator is based on the analysis of 622 of the largest reservoirs in the world (storage capacity at least 0.5 km^3). The residence time of otherwise free flowing water is termed by the authors "aging of continental freshwater."
Exploitation of groundwater resources	Compilation of data and case studies are from the following sources: EEA 1995, British Geological Survey 1996, Foster et al. 1998, and Scheidleder et al. 1999.
Wetlands extent and change in the United States and estimates for some European countries	Data for the United States are from the National Wetlands Inventory, U.S. Fish and Wildlife Service (USFWS), and the Natural Resource Inventory of the U.S. Department of Agriculture. European data are from the European Environment Agency.
Percentage of cropland and urban and industrial land use by river basin	Cropland area is estimated from the Global Land Cover Characterization Database (GLCCD 1998) at 1-km resolution. Cropland in this analysis excludes areas of mixed natural/cropland vegetation. Urban/industrial areas are based on NOAA-NGDC's Stable Lights and Radiance Calibrated Lights of the World CD-ROM (1998). The data set contains the locations of stable lights, including frequently observed light sources, such as gas flares at oil drilling sites. Data were collected in 1994–95.

CONDITIONS AND TRENDS

- ♦ Although water in rivers, lakes, and wetlands contains only 0.01 percent of the world's freshwater and occupies less than 1 percent of the Earth's surface, the global value of freshwater services is estimated in the trillions of U.S. dollars.
- ♦ Dams have a significant impact on freshwater ecosystems. Large dams have increased sevenfold in number since 1950 and now impound 14 percent of the world's runoff.
- ♦ Sixty percent of the largest 227 rivers of the world are strongly or moderately fragmented by dams, diversions, and canals. In all, strongly or moderately fragmented systems account for nearly 90 percent of the total water volume flowing through these rivers.
- ♦ In the developing world, large dams are still being built at a fast rate, threatening the integrity of some of the remaining free-flowing rivers in the world. The basins with the greatest number of large dams currently under construction are the Yangtze, the Tigris and Euphrates, and the Danube.
- ♦ According to estimates by Vörösmarty et al. (1997a and 1997b), the average residence time of river water in regulated basins has tripled to over one month worldwide, whereas large reservoirs trap 30 percent of the global suspended sediments.
- ♦ Half the world's wetlands are estimated to have been lost during the 20th century, as land was converted to agriculture and urban use, or filled to combat diseases, such as malaria.
- ♦ At least 1.5 billion people rely on groundwater as their only source of drinking water. Overexploitation and pollution in many regions of the world are threatening groundwater supplies, but comprehensive data on the quality and quantity of this resource are not available at the global level.

INFORMATION STATUS AND NEEDS

- ♦ Global information on dams and reservoirs is limited to dams that are 15 meters in height or greater, except for China, Japan, India, and Spain, which report only on dams over 30 meters. The largest data gaps are for Russia, which reports only on hydropower dams, and China, where the majority of the world's large dams have been built and for which information is exceedingly difficult to acquire.
- ♦ The provision of latitude and longitude for each dam would highly improve our ability to locate these structures within the correct hydrological unit and assess their downstream impacts.
- ♦ Information on discharges is also lacking for many reservoirs; however, these data are needed to assess more fully the annual variations in river flow.
- ♦ Another important data set needed to assess freshwater ecosystem conditions is complete global information on wetlands distribution and change. Location and size of wetlands is especially needed for Asia, Africa, South America, the Pacific Islands, and Australia.
- ♦ Regional data for Oceania, Asia, Africa, eastern Europe, and the Neotropics allow for only cursory assessment of wetlands extent and location. Only North America and western Europe have better data and monitoring programs in place to track changes in wetlands area.
- ♦ Remote sensing data from the new Landsat 7 satellite and from radar, which can sense flooding underneath vegetation and can penetrate cloud cover, should improve the information base on the extent, location, and change in wetlands.
- ♦ Limited information is available on groundwater exploitation at the global level. National-level data exist but are not readily accessible or are not harmonized among countries. Groundwater information should be collected in coordination with data collection efforts on the effects of their use on other regional water resources, such as wetlands, lagoons, and river basins.

Water Quantity

PAGE Measures and Indicators	Data Sources and Comments
Annual renewable water supply per person by river basin in 1995 and projections for 2025	CIESIN et al. (2000), Global Population Database.This database is based on census data for over 120,000 subnational administrative units for 1995. Water supply estimates are from a global runoff database developed by Fekete et al. (1999) at the University of New Hampshire in collaboration with the WMO/Global Runoff Data Centre in Germany. It combines observed discharge data with modeled runoff data.
Annual renewable water supply and dry season flow by river basin	Runoff estimates are from a global runoff database developed by Fekete et al. (1999). The dry season flow is estimated by selecting the four driest consecutive months of the year for each basin.

Conditions and Trends

- Between 1900 and 1995, water withdrawals increased sixfold, more than twice the rate of population growth. Dams and reservoirs have helped provide drinking water for much of the world's population, increased agricultural output through irrigation, eased transport, and provided flood control and hydropower.
- Many regions of the world have ample water supplies, but currently more than 40 percent of the world's population live in river basins experiencing water stress.
- As the world population grows from six to nine billion by the middle of the 21st century, we will become more dependent on irrigation for our food supplies, which will exacerbate the water scarcity problem in many regions and push other regions and populations to situations of water stress.
- By 2025, the PAGE analysis projects that, assuming current consumption patterns continue, at least 3.5 billion people or 48 percent of the world's population will live in water-stressed basins.
- Based on the U.N. low-range population growth projection, 63 river basins are projected to have a population greater than 10 million by 2025. Of these river basins, 29 are already water stressed and will descend further into scarcity, 6 will move into water-stress conditions, and 12 additional basins may experience a strong negative change in water supply per person between 1995 and 2025.
- Low dry season flows have exacerbated water supply and quality problems in 27 basins with more than 10 million people in 1995. These basins include the Balsas and Grande de Santiago in Mexico, the Limpopo in Southern Africa, the Hai Ho and Hong in China, the Chao Phraya in Southeast Asia, and the Brahmani, Damodar, Godavari, Krishna, Mahi, Narmada, Ponnaiyar, Rabarmarti, and Tapti in India.

Information Status and Needs

- Statistics on water availability and use at the global scale are poor. In many parts of the world, we know less about water resources than we did 20 years ago. The number of functioning hydrological stations, for example, has fallen significantly since 1985.
- Current statistics of water withdrawals and consumption are fraught with uncertainty because of the highly decentralized nature of water use.
- Most estimates are based on a combination of modeled and observed data.
- In order to improve our ability to monitor the condition of freshwater systems to provide water for humans and ecosystems, better statistics on water availability and use are urgently needed, preferably at the watershed level so that impacts on entire ecosystems can be monitored.

Water Quality

PAGE Measures and Indicators	Data Sources and Comments
Global concentrations of biochemical oxygen demand (BOD), phosphorous, and nitrates by river basin	Data are from UNEP's Global Environmental Monitoring System (GEMS) Water Programme (1995). This project measured water quality in 82 major river basins from 1976 to 1990. Measurements are from a network of 175 sampling stations in around 60 countries. Because data from sampling points are extrapolated to the entire watershed, these data should be interpreted with caution.
Trends in phosphorous and nutrient concentrations in Europe and the United States	Data for Europe are from the European Environment Agency. Data for the United States are from the U.S. Geological Survey (USGS) National Water-Quality Assessment (NAWQA) program, and the USGS National Stream Quality Accounting Network (NASQAN). NASQAN monitors water quality in the four largest river systems in the United States: the Mississippi (including the Missouri and Ohio), the Columbia, the Colorado, and the Rio Grande. NAWQA performs detailed studies in 60 smaller basins across the United States, including Alaska and Hawaii.
Biological methods of water quality monitoring	Data are from studies from around the world, including the United States, France, India, the United Kingdom, and Australia. All studies illustrate applications of biological criteria to monitor water quality.
Nitrate pollution in groundwater	Data are from various sources and studies for China, India, western Europe, and the United States.

Conditions and Trends

- Water-borne diseases from fecal contamination of surface waters continue to be a major cause of mortality and morbidity in the developing world.
- Surface water quality has improved in the United States and western Europe in the past 20 years with respect to some pollutants; however, nutrient loading from agricultural runoff continues to be a problem in these two regions.
- Worldwide water quality conditions appear to have degraded in almost all regions with intensive agriculture and large urban/industrial areas.
- Cases of algal blooms and eutrophication are being documented more frequently in most inland water systems around the world.
- Of the 82 major river basins in the world, those in North America, Europe, and Africa had the highest concentration of organic matter for the period 1976–90.
- Phosphorous concentration in U.S. waterways show improvement, whereas nitrate concentrations have remained more or less stable for the 1980-89 period.
- Evidence shows that nitrate pollution in groundwater, from fertilizer use, is getting worse in northern China, India, and Europe. Population increases in these areas and the need to increase agricultural production will require increase use of fertilizers, which will exacerbate the groundwater pollution problem.

Information Status and Needs

- Data on water quality at the global level is very scarce. There have been very few sustained programs to monitor water quality worldwide.
- Information is usually limited to industrial countries or small, localized areas.
- Water monitoring is also almost exclusively limited to chemical pollution rather than biological monitoring, which would provide a better understanding of the condition of the system. For regions, such as Europe, where some monitoring is taking place, difference in measures and approaches make the data hard to compare.

Food Production – Inland Fisheries

PAGE MEASURES AND INDICATORS	DATA SOURCES AND COMMENTS
Historical change in fish catch and species composition for selected rivers, lakes, and inland seas.	Data are from various sources for the following bodies of water: Danube, Rhine, Missouri, Great Lakes, Illinois, Pearl (Xi Jiang), Lake Victoria, Colorado in the United States, and the Aral Sea. All studies looked at either changes in species composition or changes in commercial landings of important inland fisheries.
Recent trends in catch statistics from inland waters	Inland capture fisheries data are from the Food and Agriculture Organization of the United Nations (FAO) for the period 1984–97. Inland capture fisheries include freshwater and diadromous fish caught in inland waters, and freshwater molluscs and crustaceans.

CONDITIONS AND TRENDS

- In 1997, inland fisheries landings accounted for 7.7 million metric tons, or almost 12 percent of total capture available for human consumption, a level estimated to be at or above maximum sustainable yields. Taking into account the inland capture, fisheries are estimated to be underreported by two or three times, the contribution to direct human consumption is likely to be at least twice as high.
- Freshwater aquaculture currently has a higher production than capture fisheries, contributing 17.7 million metric tons of fish and seafood in 1997. In 1997, marine and inland aquaculture production provided 30 percent of the fish for human consumption; 60 percent of this production comprised of freshwater finfish or fish that migrate between fresh and saltwater.
- At the global level, inland fisheries landings have been increasing since 1984. Most of this increase has occurred in Asia, Africa, and more moderately in Latin America. In North America, Europe, and the former Soviet Union, landings have declined, whereas in Oceania they have remained stable.
- Despite this increase in landings, maintained in many regions by fishery enhancements, such as stocking and fish introductions, the greatest overall threat for the long-term sustainability of inland fishery resources is the loss of fishery habitat and the degradation of the terrestrial and aquatic environment.
- Historical trends in commercial fisheries data for well-studied rivers show dramatic declines over the 20th century, mainly from habitat degradation, invasive species, and overharvesting.

INFORMATION STATUS AND NEEDS

- Data on inland fisheries landings are poor, especially in developing countries. The FAO database on inland fisheries landings is the most complete data set at the global level; however, it has important limitations. Some of the main problems are that much of the catch is not reported at the species level and much of the fish consumed locally is never reported, making fishery assessment difficult.
- There is no systematic data collection on the contribution of stocking, fish introductions, and other enhancement programs to inland fisheries. This information, as well as information on recreational fisheries, which are becoming increasingly important in many countries, should be incorporated into data collection efforts.
- Reporting on fishery resources at the watershed level instead of the national level, as it has been done to date, would improve our understanding of the condition of the system and the linkages between upstream activities and their downstream effects. This information could then be applied to watershed and fishery resources management plans.
- Historical trends in fisheries statistics are available only for a few well-studied rivers, and because of the multispecies composition of the catch in most inland water bodies, particularly in developing countries, assessments on the condition of the resources are hard to carry out.

Freshwater Biodiversity

PAGE Measures and Indicators	Data Sources and Comments
Important areas and ecoregions for freshwater biodiversity	Olson and Dinerstein 1999 and Groombridge and Jenkins 1998. Both analyses are priority-setting exercises for conservation, based on existing data and expert opinion.
Fish species richness and endemism by river basin	Revenga et al. 1998. Data compiled for the World Resources Institute by the World Conservation Monitoring Centre (WCMC). Additional information comes from Kottelat and Whitten (1996) and Oberdorff (1997).
Biological distinctiveness index for North America	Abell et al. 2000. Regional priority-setting analysis for conservation, based on a combination of existing environmental data and expert opinion.
Bird population trends in the United States and Canada	Data are from the North American Breeding Bird Survey, which is organized by the Patuxent Environmental Science Center. Data used in this report are limited to wetland-dependent species. Population trends cover the period 1966–98.
Global amphibian population census	Data are from the Declining Amphibian Populations Task Force (DAPTF). DAPTF is a network of more than 3,000 scientists working in 90 countries.
Threatened species and habitats in North America, the Middle East, and Europe	Data for North America are from Abell et al. 2000. These data cover threat status for North American fish and reptile species. Data for Europe and the Middle East are from BirdLife International. Data are for threatened bird species and important bird areas in these two regions.
Presence of nonnative species: introduced fish, zebra mussel in the United States, and water hyacinth distribution	Introduced fish species information is from FAO's Database on Introductions of Aquatic Species (DIAS). Data on zebra mussel expansion are from the USGS Non-indigenous Aquatic Species (NAS) information resource. Data on global distribution of water hyacinth are from a variety of sources. Water hyacinth distrubution in the United States is from the Aquatic Nuisance Species Task Force, cochaired by the USFWS and the National Oceanic and Atmospheric Administration (NOAA).

Conditions and Trends

- Freshwater ecosystems harbor an extraordinary concentration of species; approximately 300 new freshwater species are described each year. World Wildlife Fund-US (WWF-US) has identified 53 freshwater ecoregions around the world as priority areas for conservation, based on their unique assemblage of species, habitats, and ecological or evolutionary phenomena, while the WCMC has identified 136 areas of high freshwater biodiversity around the world.
- Physical alteration, habitat loss and degradation, water withdrawal, pollution, overexploitation, and the introduction of nonnative species all contribute to declines in freshwater species.
- More than 20 percent of the world's freshwater fish have become extinct or been threatened or endangered in recent decades.
- Of the 108 large basins analyzed, 27 have high fish species richness. More than half of these basins are in the tropics, and the rest are in central North America, India, and China.
- Evidence shows that freshwater species, such as amphibians, fish, and wetland-dependent birds, are at high risk of imperilment in many regions of the world. In the United States and parts of Canada, however, 66 percent of the populations of wetland birds are increasing.
- The intentional or accidental introduction of nonnative species in freshwater systems is a global phenomenon. Evidence for North America, one of the best-documented regions, shows that the introduction of alien species not only has contributed to the extinction and imperilment of native fauna but also has substantial associated economic costs.
- Modeled estimates of future species extinction rates suggest that the rates for freshwater animal species are five times higher than for terrestrial species.
- The growing concern for species, the maintenance of pristine habitats, and the need to maintain other goods and services, such as clean water, is driving the trend, in some countries, to restore and rehabilitate freshwater systems.

Information Status and Needs

- Direct measurements of the condition of biodiversity in freshwater systems are sparse worldwide. Basic information on freshwater species for many developed nations and most of the developing world is lacking, as well as threat-analyses for most freshwater species. This makes analyzing population trends impossible or limited to a handful of well-known species.
- Information on nonnative species is frequently anecdotal and often limited to records of the presence of a particular species, without documentation of the effects on the native fauna and flora. Spatial data on invasive species are available for few species, mostly in the United States and Australia.
- Excellent trend data are available for bird populations in the United States and Canada, and other available regional data are good but lack long-term population trends, such as data on the distribution of important bird areas from BirdLife International.
- At a minimum, there should be monitoring of key indicator species and monitoring of the presence or introduction of nonnative species and their impacts on native fauna and flora.

Conclusions

Human activities have severely affected the condition of freshwater systems worldwide. Even though humans have increased the amount of water available for use with dams and reservoirs, more than 40 percent of the world's population lives in conditions of water stress. This percentage is estimated to grow to almost 50 percent by 2025. Surface and groundwater is being degraded in almost all regions of the world by intensive agriculture and rapid urbanization, aggravating the water scarcity problem. In addition, lack of access to clean water continues to be a leading cause of illness and death in much of the developing world. Food production from wild fisheries has been affected by habitat degradation, overexploitation, and pollution to a point where most of these resources are not sustainable without fishery enhancements. Finally, the capacity of freshwater ecosystems to support biodiversity is highly degraded at a global level, with many freshwater species facing rapid population declines or extinction.

The PAGE study relied on existing global and regional data collected by organizations and scientists around the world. Without their efforts and their willingness to share the data, this study would not have been possible. We found many detailed data sets at the national level, particularly for the United States. However, at the global level, data on freshwater resources are scarce. Basic statistics on water availability and use, for example, are not readily available or are available only at the national level. This makes an assessment of freshwater systems difficult because the ideal biophysical unit of analysis is the watershed—which often crosses national boundaries.

Improved national and global data on ecosystem land use characteristics, basic hydrological information, fisheries production, and freshwater species could lead to significantly more knowledge about the condition of freshwater systems. Better information on actual stream and river discharge, and the amount of water withdrawn and consumed, would increase our ability to manage freshwater systems more efficiently and evaluate trade-offs. But much effort and financial commitment would have to be made to restore many of the hydrological stations around the world, which have been declining since the mid-1980s and, in some cases, are no longer functioning.

Remotely sensed data from new satellites with higher quality sensors and much larger onboard storage capacity will allow better analysis of the changes in land-use patterns, and may allow for a complete and accurate mapping of the extent of both seasonal and permanent wetlands. With this information, resource managers would be able to use existing models to better understand how watershed changes are likely to affect the water quantity, and quality, of rivers and lakes.

In terms of food production from inland waters, there is an urgent need to improve the quality of the data on inland capture fisheries and those environmental and socioeconomic factors that affect their sustainability. Improved inland fishery data are likely to require improved or new monitoring networks, which means a financial commitment to strengthen both national fisheries departments and collaborating organizations, such as the Food and Agriculture Organization of the United Nations.

Better information on water quality can provide nations with immediate benefits because of the direct connection between water quality and human health. But gathering such information generally requires expensive monitoring networks that are beyond the reach of many developing countries. Even though surface water monitoring programs are well developed in most Organization for Economic Cooperation and Development (OECD) countries, water quality monitoring in most parts of the world is rudimentary or nonexistent. Even those developed countries that have water quality monitoring programs in place focus on chemical parameters that leave out important biological information. One of the biggest challenges in future water monitoring programs is the integration of chemical and biological measures of water quality.

If surface water monitoring is still deficient in many countries, the situation for groundwater is worse. Many nations lack proper monitoring of groundwater recharge quality. Information on groundwater quality, as well as on aquifer storage capacity and exploitation, is urgently needed. Currently there are two proposed initiatives that could help fill in the information gap on groundwater resources as well as promote their sound management. The first is a high-level Groundwater Management Advisory Team, coordinated by the World Bank and the Global Water Partnership. The aim of this team is to promote more effective management of groundwater resources around the world, through improved understanding of the hydrogeological constraints and strengthening of the institutional framework. The second is an International Groundwater Resources Assessment Center, that will collect data on and monitor groundwater worldwide. This initiative is being coordinated by UNESCO and the World Meteorological Organization.

Finally, as this study will stress, information on freshwater biodiversity is poor even in those developed nations that have considerable financial and technical resources. There are several new international initiatives, including the OECD's Global Biodiversity Information Facility (GBIF) and the Global Taxonomy Initiative of the Convention on Biological Diversity that will help identify and catalogue species around the world. This knowledge and monitoring would allow for a more complete assessment of the condition of freshwater systems.

Recommendations for the Millennium Ecosystem Assessment include the following:

- Compiling more complete information on the precise location of the world's dams, including the thousands of dams less than 15 meters in height that are not currently listed in international databanks.
- Developing a complete global spatial data set on wetlands distribution. Information on the location and size of wetlands is especially needed for Asia, Africa, South America, and Oceania.
- Promoting the restoration of the hydrological monitoring stations and improving the water supply and use statistics at the watershed level.
- Compiling a global data set on groundwater resources, including their distribution, capacity, and use.
- Encouraging national governments to establish water quality monitoring programs that combine chemical and biological measures for both surface and groundwater.
- Strengthening the role of national fisheries departments in data collection for inland fisheries resources.
- Establishing a systematic data collection effort on the contributions of fish stocking, introduction, and other enhancement programs to inland fisheries.
- Monitoring key indicator species for freshwater systems, as well as monitoring the presence or introduction of nonnative species and their impacts on native fauna and flora.

Prologue: Freshwater Systems

What They Are, Why They Matter

Freshwater ecosystems in rivers, lakes, and wetlands contain just a fraction—one-hundredth of 1 percent—of the Earth's water and occupy less than 1 percent of the Earth's surface (Watson et al. 1996:329; McAllister et al. 1997:18). Yet these vital systems render services of enormous global value that are on the order of several trillion U.S. dollars, according to some estimates (Postel and Carpenter 1997:210).

The most important goods and services that humans derive from freshwater systems revolve around water supply: providing a sufficient quantity of water for domestic, agriculture, and industrial use; maintaining high water quality; and recharging aquifers that feed groundwater supplies. But freshwater ecosystems provide many other crucial goods and services as well: fish for food and sport, biodiversity, mitigation of floods, assimilation and dilution of wastes, nutrient cycling and restoration of soil fertility, recreational opportunities, aesthetic values, and transportation for both people and goods. Harnessed by dams, these systems also produce hydropower, one of the world's most important renewable energy sources.

Prior to the 20th century, global demand for these goods and services was small compared to what freshwater ecosystems could provide. Historically, human development has favored activities with high economic returns that often maximized single objectives, such as abstracting water for irrigation schemes and urban aqueducts, draining mosquito-ridden swamps, and erecting dams to produce electricity and control seasonal water flows. Conflicts between different users and uses were more frequent in areas with an unreliable water supply. In areas with more reliable runoff patterns, the biggest problems were flood hazards and localized pollution problems when the discharge of human effluent exceeded the waste assimilation capacity of freshwater systems.

With population growth, industrialization, and the expansion of irrigated agriculture, demand for all water-related goods and services has increased dramatically, straining the capacity of freshwater systems. Although many policymakers are aware of the growing problems of water scarcity, there are many other signs of freshwater stress. The total amount and the number of pollutants entering freshwater systems, for example, has grown from a limited amount of organic matter from human and animal wastes and a few metals from mining to large quantities of human effluent and thousands of chemicals, such as pesticides and fertilizers.

The number of large dams (over 15 meters high) has increased sevenfold since 1950, from about 5,750 to more than 41,000 (ICOLD 1998:7, 13), impounding at least 14 percent of the world's annual runoff (L'vovich and White 1990:239). Between 1950 and 2000, annual water availability per person decreased from 16,800 m^3/year to 6,800 m^3/year, calculated on a global basis and assuming 42,700 km^3/yr of global freshwater runoff (Shiklomanov 1997:73). Runoff is defined as the renewable supply of water that flows through the world's rivers after evaporation and infiltration (WMO 1997:7). Water availability per person, therefore, refers to the amount of this renewable supply of water divided by the global population. With global population expected to reach at least 7.8 billion by 2025—the U.N. medium population projection (UNPD 1999:2)—per capita water availability is estimated to fall to 5,400 m^3/year.

As the amount of water available on a per capita basis declines, trade-offs between alternative water uses become more acute in terms of the environmental implications.

In many rivers, ecosystem functions or responses have been lost or impaired to the point at which human values and species diversity are adversely affected and restoration or protection is necessary to sustain natural watershed services. Maximizing one environmental good is no longer possible without significant trade-offs for other goods and services. Resource competition and conflicts are growing, becoming regional and global in scale. Managing freshwater systems increasingly will require integrating multiple objectives and data, using a basin and ecosystem approach to comprehensively assess the impacts on biological, chemical, and hydrological systems. Such an approach is especially important because some key freshwater services, such as water purification, maintenance of biodiversity, and watershed protection, never enter the market and, thus, have no price tag. This makes it harder to assess the trade-offs at stake when different uses of a freshwater resource are proposed.

The example of Africa's Lake Victoria illustrates the profound and unpredictable trade-offs that can occur when management decisions are made without regard to the ecosystem's reaction. In Lake Victoria, maximizing one particular good in concert with increasing resource pressure has caused drastic ecological changes. It also has led to a shift in the distribution of economic benefits from the previously large number of local beneficiaries who obtained a livelihood at a very modest level from the fisheries to a few who could afford to invest or participate in international fish exports.

Lake Victoria, bounded by Uganda, Tanzania, and Kenya, is the world's largest tropical lake, and its fish are an important source of food and employment for the region's 30 million people. Before the 1970s, Lake Victoria contained more than 350 species of fish in the cichlid family, of which 90 percent were endemic, giving it one of the most diverse and unique assemblages of fish in the world (Kaufman 1992:846–847, 851). Today, more than half of these species are either extinct or found only in very small populations (Witte et al. 1992:1, 17).

The collapse in the lake's biodiversity was caused primarily by the introduction of two exotic fish species, the Nile perch and Nile tilapia, which fed on and outcompeted the cichlids for food. But other pressures factored in the collapse as well. Overfishing depleted native fish stocks and provided the original motivation for introducing the Nile perch and tilapia in the early 1950s. Land-use changes in the watershed dumped pollution and silt into the lake, increasing nutrient load and causing algal blooms and low oxygen levels in deeper waters—a process called eutrophication. These changes resulted in major shifts in the lake's fish populations. Cichlids once accounted for more than 80 percent of Lake Victoria's biomass and provided much

Figure 1

Trading Biodiversity for Export Earnings: The Changing Lake Victoria Fishery

(Kenya only)

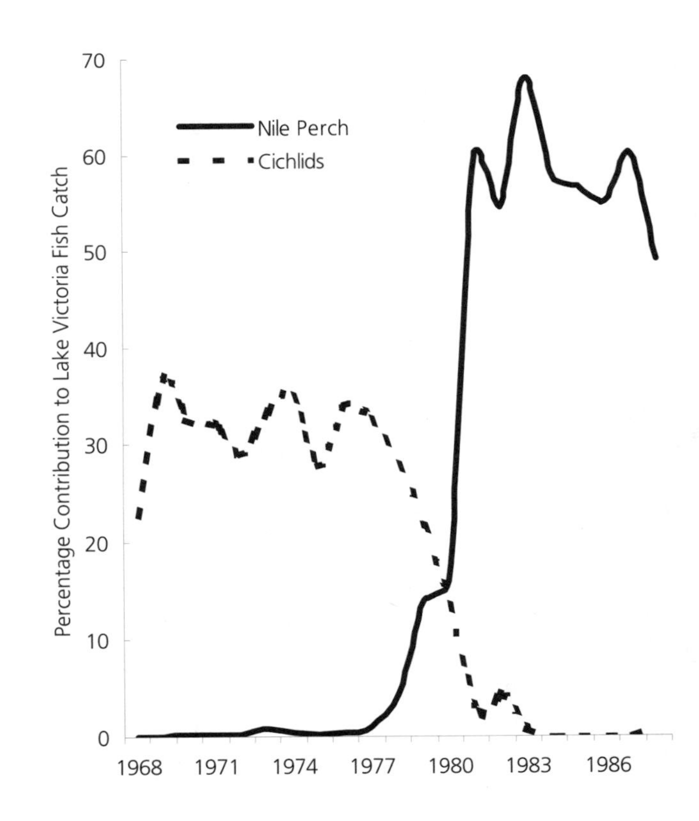

Source: Achieng 1990.

of the fish catch (Kaufman 1992:849). By 1983, Nile perch made up almost 70 percent of the catch, with Nile tilapia and a native species of sardine making up most of the balance (*see Figure 1*) (Achieng 1990:20).

Although the introduced fish devastated the lake's biodiversity, they did not destroy the commercial fishery. In fact, total fish production and its economic value rose considerably. The Nile perch fishery now produces some 300,000 metric tons of fish yearly (FAO 1998), earning US$280 million to US$400 million in the export market—a market that did not even exist before the perch was introduced (Kaufman, personal communication, 2000). Unfortunately, local communities that had depended on the native fish for decades did not benefit from the success of the Nile perch fishery, primarily because Nile perch and tilapia are caught with gear that local fishermen could not afford. And, because most of the Nile perch and tilapia are shipped out of the region, the local availability of fish for consumption has declined. In fact, while tons of perch find their way to restaurants as far away as Israel and Europe, there is evidence of protein malnutrition among the people of the lake basin (Kaufman, personal communication, 2000).

The sustainability of the Nile perch fishery is also a concern. Recent evidence suggests that eutrophication and oxygen depletion in the lake, as well as overfishing in certain areas, are already threatening the long-term sustainability of the Nile perch fishery. The stability of the entire aquatic ecosystem—so radically altered over a 20-year span—is in doubt (Kaufman 1992:850). The ramifications of the species introductions can even be seen in the watershed surrounding Lake Victoria. Drying the perch's oily flesh to preserve it requires firewood, unlike the cichlids, which could be air-dried. This has increased pressure on the area's limited forests, increasing siltation and eutrophication, which, in turn, has further unbalanced the precarious lake ecosystem (Kaufman 1992:849–851; Kaufman, personal communication, 2000).

In sum, introducing Nile perch and tilapia to Lake Victoria traded the lake's biodiversity and an important local food source for a significant—although perhaps unsustainable—source of export earnings. When fisheries managers introduced these species, they unknowingly altered the balance of goods and services the lake produced and redistributed the economic benefits flowing from them. Knowing the full dimensions of these trade-offs, would they make the same decision today?

Another very different case that illustrates these trade-offs between environmental goods and services over time and the cost to recover some of these services is the Skjern River floodplain restoration effort in Denmark. The Skjern River is the largest river in Denmark, with a watershed area covering 6 percent of the country. It is the main source of freshwater and nutrients to the Ringkøbing Fjord, which is a coastal, shallow lagoon connected to the North Sea (NFNA 1997:5). The downstream floodplain, where the restoration project is taking place, occupies about 1 percent of the watershed (Riber, personal communication, 2000). Agriculture with extensive animal husbandry is the predominant land use in the watershed. The course of the lower Skjern River has been modified several times since the 18th century, with the greatest change taking place in the 1960s when the lower 20 kilometers of the river were straightened and confined within embankments. This last modification converted 4,000 hectares of wetlands, meadows, and marshlands into farmland, mostly for grain production, reducing the wetland area to only 2 percent of its original extent (*see Figure 2*) (NFNA 1999:6, 7).

One benefit of the river channelization, in addition to providing more land for agriculture, was to reduce the frequency of floods. Although flooding of the floodplain was prevented, a

Figure 2

Skjern River Floodplain: Marshland and Meadow Area in 1871 and 1987

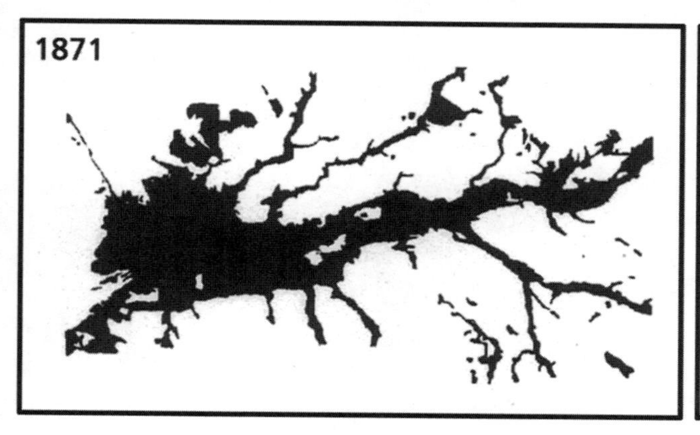

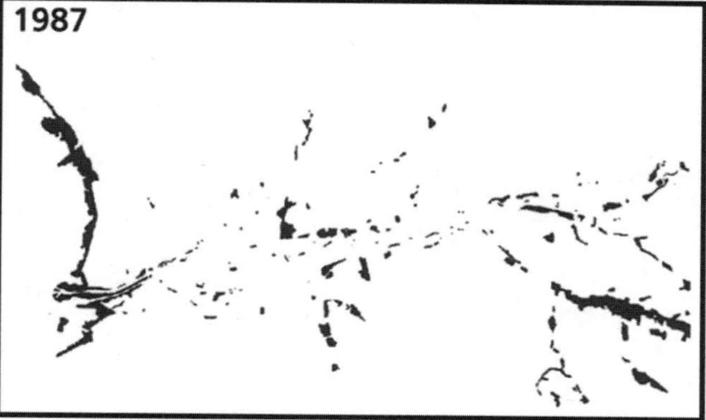

Source: DHI Water and Environment 1999.

new flooding risk in the neighboring towns emerged, but these flooding events were marginal (Riber, personal communication, 2000). Also on the negative side of the ledger, channel modification reduced biodiversity in the watershed. For example, otters disappeared, nesting waterfowl declined, and the salmon population—the last wild salmon population in Denmark—was reduced to a small fraction of its peak (NFNA 1999:4, 24, 25). Draining wetlands caused other environmental changes, including ground subsidence of up to one meter in certain areas, ochre leaching, and waterlogging that reduced the productivity of the reclaimed agricultural land. In addition, the intensive agricultural use brought higher nutrient loadings to the fjord, contributing to its eutrophication (Olesen and Havnø 1998:3; NFNA 1999:12, 13).

The Skjern River floodplain restoration project was motivated by changes in societal values and priorities, notably the decreased economic importance of agriculture in the country, an increased appreciation of nature for recreation and tourism, and the desire to rectify past environmental damage. Work began in 1999 with the aim of restoring the Skjern River to a "natural" river by eliminating embankments, returning the river to its original meandering course, and recreating wetlands. The goal is to bring back salmon, bird, and plant populations. Furthermore, the new wetlands will act as filters, decreasing eutrophication in the fjord and helping to restore its biodiversity. The cost of the restoration project is estimated at US$35 million. It will take 3 years to restore the planned 2,200 hectares of wetlands. The area will be operated as a natural park by the National Forest and Nature Agency, which is also responsible for implementing the project (NFNA 1999).

Restoration projects like the Skjern River floodplain incorporate, albeit retroactively, all the elements that are important for the management of freshwater systems. They rely on scientific data and modeling, integrate multiple objectives, use a basin and ecosystem approach, and look at trade-offs between different goods and services. To avoid costly restoration projects, future assessments of freshwater systems need to include as many of these elements as possible. The level of detail and comprehensiveness will vary with the scale and purpose of the assessment. Assessments at a global level that are used to set priorities and identify key trends will require less detail, but they will ultimately constitute a much needed worldwide integration of more apparent regional patterns.

Human Modification of Freshwater Systems

Freshwater systems have been altered since historical times, but such modifications skyrocketed in the early to mid-1900s. Table 1 shows a subset of these alterations. The projects include modifying waterways to improve navigation, draining wetlands, constructing dams and irrigation channels, and establishing interbasin connections and water transfers. These changes have improved transportation, provided flood control and hydropower, and boosted agricultural output by making more land and irrigation water available. At the same time, these physical changes in the hydrological cycle disconnect rivers from their floodplains and wetlands and slow water velocity in riverine systems, converting them to a chain of connected reservoirs. This, in turn, impacts the migratory patterns of fish species and the composition of riparian habitat, opens up paths for exotic species, changes coastal ecosystems, and contributes to an overall loss of freshwater biodiversity and inland fishery resources. Dams, on average, also affect the seasonal flow and sediment transport of rivers for 100 kilometers downstream. Some major water projects, such as the Aswan High Dam in Egypt, have an effect that extends more than 1,000 kilometers downstream (McAllister et al. 1997:56).

The PAGE analysis looks at the following indicators of modifications to freshwater systems:

- Indicators of modification of rivers: river fragmentation and flow regulation, sediment, and nutrient retention.
- Indicators of modification of groundwater resources: overview of overexploitation and saltwater intrusion problems around the world.
- Indicators of modification of wetland area: loss of wetlands in the United States and Europe.
- Indicators of modification at the watershed level: cropland and urban or industrial land use by watershed.

Modification of Rivers

Humans have built large numbers of dams all over the world, most of them in the last 35 years. Today, there are more than 40,000 large dams (more than 15 meters high) in the world, 21,600 of which are in China alone (ICOLD 1998:7; IJHD 1998:60). This storage capacity represents a 700 percent increase in the standing stock of water in river systems compared to natural river channels since 1950 (Vörösmarty et al.

Table 1
Alteration of Freshwater Systems Worldwide

Alteration	Pre-1900	1900	1950–60	1985	1996–98
Waterways Altered for Navigation	3,125 km	8,750 km	–	>500,000 km	–
Canals	8,750 km	21,250 km	–	63,125 km	–
Large Reservoirs*					
Number	41	581	1,105	2,768	2,836
Volume (km^3)	14	533	1,686	5,879	6,385
Large Dams (>15 m High)	–	–	5,749	–	41,413
Installed Hydro Capacity (MW)	–	–	< 290,000	542,000	~660,000
Hydro Capacity Under Construction (MW)	–	–	–	–	~126,000
Water Withdrawals	–	578 km^3/year	1,984 km^3/year	~3,200 km^3/year	~3,800 km^3/year
Wetlands Drainage**	–	–	–	160,600 km^2	–

*Large reservoirs are those with a total volume of 0.1 km^3 or more. This is only a subset of the world's reservoirs.
**Includes available information for drainage of natural bogs and low-lying grasslands as well disposal of excess water from irrigated fields. There is no comprehensive data for wetland loss for the world.
Sources: Based on Naiman et al. 1995, as adapted from L'vovich and White 1990. Data on dams are from ICOLD 1998. Reservoir data are from Avakyan and Iakovleva 1998. Hydro capacity data are from IJHD 1998 and L'vovich and White 1990. Water withdrawal data are from Shiklomanov 1997.

1997a:210). Table 2 shows the distribution of large dams by continent, based on the 25,410 registered dams reported by the International Commission on Large Dams (ICOLD) and the storage capacity of large reservoirs.

In terms of storage capacity, Asia and South America have seen the biggest recent increase in the number of reservoirs. In Asia, 78 percent of the total reservoir volume was constructed in the last decade, and in South America almost 60 percent of all reservoirs have been built since the 1980s (Avakyan and Iakovleva 1998:47). The inventory of dams and reservoirs is incomplete for China and the former Soviet Union, which with the United States are the world's top ranking countries in terms of number of large dams (ICOLD 1998:7). Reservoirs with more than 0.5 km^3 maximum storage capacity intercept and trap an estimated 30 percent of global suspended sediments (Vörösmarty et al.1997b:271).

Dams provide a variety of important benefits to society. They store water for agricultural, domestic, and industrial use, provide flood control and hydropower, and create reservoirs and downstream releases that society uses for recreational purposes. At the same time, dams can cause severe environmental impacts. These environmental impacts include interference with fish migratory routes, destruction of riparian habitat and species breeding grounds (through flooding of riparian habitat and the loss of wetlands and sandbanks), siltation of reservoirs, effects in coastal areas and deltas resulting from increased sedimentation and nutrient loads, changes in the temperature and chemical composition of the dammed river, decline in water availability through evaporation from reservoirs, and an increase in the occurrence of algal blooms and water-related diseases in reservoirs (Ward and Stanford 1989:56–62; Ligon et al. 1995:185–188; Vörösmarty et al. 1997a:216–218; McCully 1996:30–48).

RIVER CHANNEL FRAGMENTATION AND FLOW REGULATION

River fragmentation, which is the interruption of a river's natural flow by dams, inter-basin transfers, or water withdrawal, is an indicator of the degree that rivers have been modified by humans (Ward and Stanford 1989; Dynesius and Nilsson 1994). This indicator provides insight on the condition of these river systems and their capacity to provide a variety of goods and services.

This study assesses most of the world's large rivers to quantify the extent to which dams and canals have fragmented river basins and to determine how water withdrawals have altered river flows. The measures used to assess fragmentation and regulation include dams, reservoirs, interbasin transfers, and irrigation consumption. Irrigation consumption refers to the water that is evaporated or used by crops through transpiration, but excludes the amount of water returned to the river after irrigation.

Table 2

Large Dams and Storage Capacity of Large Reservoirs by Continent

Continent	World Registered Dams		Number of Large Reservoirs*	Storage Capacity of Large Reservoirs* Total Volume (km³)
	*Number***	*Percent*		
Africa	1,265	5	176	1,000
Asia	8,485	33.4	815	1,980
Oceania	685	2.7	89***	95***
Europe	6,200	24.4	576	645
North America	7,775	30.6	915	1,692
Central and South America	1,005	3.97	265	972
Total Registered Number	25, 410**	—	—	—
Estimated Number	41,413	—	2,836	6,385

*Large reservoirs are those with a total volume of 0.1 km³ or more. This is only a subset of the world's reservoirs.
**Total number of dams may not add up due to rounding. ICOLD reports a total of 25,410 registered dams.
***Includes only Australia and New Zealand.
Sources: ICOLD 1998; Avakyan and Iakovleva 1998.

The analysis presented here builds on an earlier fragmentation analysis carried out by Dynesius and Nilsson (1994). Their analysis shows that 77 percent of the total water discharge of the 139 largest river systems in the northern third of the world (North America, Europe, and the former Soviet Union) is strongly to moderately affected by fragmentation of the river channels. Large river systems (LRS) are defined as rivers with a virgin mean annual discharge (VMAD) equal to or above 350 m³ per second. The PAGE study commissioned Nilsson et al. (2000) to extend the fragmentation analysis to cover Africa, Latin America, China, and mainland Southeast Asia.

Map 1 shows the results of both analyses. Of the 227 major river basins assessed, 37 percent are strongly affected by fragmentation and altered flows, 23 percent are moderately affected, and 40 percent are unaffected. Strongly affected systems include those with less than one quarter of their main channel left without dams, where the largest tributary has at least one dam, as well as rivers whose annual flow patterns have changed substantially. Unaffected rivers are those without dams in the main channel of the river and, if tributaries have been dammed, river discharge has declined or been contained in reservoirs by no more than 2 percent.

In all, strongly or moderately fragmented systems account for nearly 90 percent of the total water volume flowing through the rivers in the analysis. All river systems with parts of their basins in arid areas or that have internal drainage systems are strongly affected. The only remaining large free-flowing rivers in the world are found in the tundra regions of North America and Russia, and in smaller coastal basins in Africa and Latin America (*see Map 1*). It should be noted, however, that considerable parts of some of the large rivers in the tropics, such as the Amazon, the Orinoco, and the Congo, would be classified as unaffected rivers if an analysis at the subbasin level were done. The Yangtze River in China, which currently is classified as moderately affected, will become strongly affected once the Three Gorges dam is completed. In the 1994 study, Dynesius and Nilsson also looked at the fragmentation of 59 medium-sized river systems (VMAD of at least 40 m³ per second but less than 350 m³ per second) in Scandinavia and found that only nine are still free-flowing.

Even though dam construction has greatly slowed in most developed countries (and some countries, such as the United States, are even decommissioning a few dams), the demand and untapped potential for dams is still high in the developing world, particularly in Asia. As of 1998, there were 349 dams over 60 meters high under construction around the world (*see Map 2*) (IJHD 1998:12–14). The countries with the largest number of dams under construction were Turkey, China, Japan, Iraq, Iran, Greece, Romania, and Spain, as well as the Paraná basin in South America. The river basins with the most large dams under construction were the Yangtze in China, with 38 dams under construction, the Tigris and Euphrates with 19, and the Danube with 11.

Waterfalls, rapids, riparian vegetation, and wetlands are some of the habitats that disappear when rivers are regulated or impounded (Dynesius and Nilsson 1994:759). These habitats are essential feeding and breeding areas for many aquatic and terrestrial species, and they also contribute significantly to maintaining other ecosystem services, such as water purification. The fragmentation indicators presented earlier indicate that many types of river ecosystems have been lost or are at a high risk, and, therefore, the populations of many river-dwelling species may have already disappeared or become highly fragmented.

SEDIMENT AND NUTRIENT RETENTION

By slowing the movement of water, dams also prevent the natural downstream movement of large amounts of sediment to deltas, estuaries, flooded forests, wetlands, and inland seas. This retention can rob the downstream areas of the sediments and nutrients that they depend on, affecting species composition and productivity. Coastal fisheries, for example, depend on upstream inputs to replenish nutrients. After the Aswan High Dam was built on the Nile River, the supply of phosphate and silicate to the coastal area was reduced to 4 and 18 percent, respectively, of pre-dam conditions (FAO 1995a:30). This drop in nutrients, combined with increased salinity in the delta because of a reduction in the Nile outflow as well as overfishing, reduced the productivity of the coastal fisheries (FAO 1995a:38, 39).

Water retention alters a river's flow regime, which eliminates or reduces spring runoff or flood pulses that often play a critical role in maintaining downstream riparian and wetland communities. These habitats are essential as breeding and feeding grounds for many fish and bird species. When dams interrupt these pulses, these habitats and associated species are lost (Abramovitz 1996:11). Sediment retention also interferes with dam operations and shortens their useful lives. In the United States, about 2 km^3 of reservoir storage capacity is lost from sediment retention each year, at a cost of US$819 million annually (Vörösmarty et al. 1997a:217).

Finally, water and sediment retention affect water quality and the waste processing capacity of rivers (the ability to break down organic pollutants). The slow-moving water in reservoirs is stratified into layers instead of being well-mixed, with the bottom waters of the reservoir often depleted of oxygen. These oxygen-starved waters can produce toxic hydrogen sulfide gas that degrades water quality. In addition, oxygen-depleted waters released from dams have a reduced capacity to process waste for up to 100 kilometers downstream because the waste-processing ability of river water depends directly on its level of dissolved oxygen.

An indicator of the extent to which dams have affected water storage and sediment retention at the global level is the change in "residence time" of otherwise free flowing water. This indicator, developed by Vörösmarty et al. (1997a), is based on the analysis of 622 of the world's largest reservoirs —defined as reservoirs with a storage capacity equal or above 0.5 km^3—and examined the increase in the time it takes an average drop of water entering the river to reach the sea. This residence time is termed by the authors as the "aging of continental freshwater."

Map 3 shows the changes in residence time or aging of river water at the mouth of each of the 236 regulated drainage basins that were analyzed (Vörösmarty et al. 1997a:210–219). Worldwide, the average age of river water in regulated basins has tripled to well over one month (Vörösmarty et al. 1997a:215). Among the basins most affected are the Colorado and Rio Grande in North America, the Nile and the Volta in Africa, and the Rio Negro in Argentina.

Changes in Groundwater Resources

Groundwater is a resource of global importance, particularly in arid and semiarid areas of the world where access to surface water is limited. Groundwater resources include shallow and deep rechargeable aquifers that are connected to rivers, streams, or seas, and nonrenewable aquifers or fossil water that may have been created by precipitation during the last Ice Age. Most aquifers are replenished by rain that infiltrates through the soil or by river losses during floods (UNEP 1996:6; Shiklomanov 1997:19). Fossil water resources, on the other hand, are not naturally or artificially recharged, so once they have been exploited they may never be replenished (UNEP 1996:7; Shiklomanov 1997:23).

It is estimated that at least 1.5 billion people use groundwater as their sole source of drinking water (UNEP 1996:4). Groundwater is also important for irrigation. For instance, more than 50 percent of the water used in India for irrigation comes from groundwater resources (Foster et al. 2000:2). Growing populations, expanding urbanization and industrialization, and increasing demands for food security are placing more pressure on the world's groundwater supplies.

There are two major consequences of the increasing need for world groundwater supplies. One is "groundwater mining," in which groundwater abstraction exceeds the natural rate of replenishment. This can result in land subsidence (in which the land sinks), saltwater intrusion, and groundwater supplies becoming economically and technically unfeasible for use as a stable water supply (UNEP 1996:4, 15). The second major consequence is the degradation of water quality resulting from a variety of point and nonpoint source pollutants, including agricultural runoff, sewage from urban centers, and industrial effluents (*see Section on Water Quality*).

We lack basic information at the global level on the size, recharge rates, and condition of groundwater supplies. There is data on distribution, but they are dispersed among national agencies and, in most cases, not harmonized (Foster, personal communication, 2000). However, in areas where groundwater is important for domestic use and irrigation, countries have produced groundwater inventories and are monitoring groundwater quality. For example, in Europe the European Environment Agency shows that nearly 60 percent of the cities with more than 100,000 people are located in areas where there is groundwater overabstraction (EEA 1995:66). Groundwater overexploitation is also evident in many Asian cities. The cities of Bangkok, Manila, Tianjin, Beijing, Madras, Shanghai, and Xian, for example, have all registered a decline in water table levels of 10–50 meters (Foster et al. 1998:23). This overexploitation in many cases is accompanied by water quality degradation and land subsidence. For instance in Mexico, where many cities have experienced declines in groundwater levels, the aquifer that supplies much of Mexico City had fallen by 10 meters as of 1992 with a consequent land subsidence of up to 9 meters (Foster et al. 1998:23, 25). One of the worst cases of groundwater overexploitation is Yemen where, in some areas, the rate of abstraction is 400 percent greater than the rate of recharge (Briscoe 1999).

In coastal zones, groundwater overabstraction can reverse the natural flow of groundwater into the ocean, causing saltwater to intrude into inland aquifers. Where coastal aquifers are used to supply drinking water or irrigation, saltwater intrusion is a serious problem. Because of the high marine salt content, a concentration of only 2 percent seawater in an aquifer is enough to make groundwater supplies unusable for human consumption (Scheidleder et al. 1999:91). A recent study of groundwater resources in Europe shows that saltwater intrusion as a consequence of overabstraction is most prevalent in the Mediterranean countries, particularly along the coastlines of Spain, Italy, and Turkey (Scheidleder et al. 1999:91). Of 126 groundwater areas, for which status was reported, 53 showed saltwater intrusion. Most of the aquifers were used for public and industrial water supply (Scheidleder et al. 1999:90, 92).

Saltwater intrusion can also occur in inland areas where groundwater overexploitation leads to the rise of highly mineralized water from deeper aquifers. This problem has been reported in Europe for 1 area in Latvia, 3 areas in Poland, and 14 in the Republic of Moldova (Scheidleder et al. 1999:90, 91).

It should be noted that data provided by different European countries are based on groundwater sampling units that are not standardized and range in size from single sampling sites to regional administrative units. The quality of data, therefore, is uneven, and it gives an incomplete picture of the situation. Countries that have not reported data are not necessarily free of groundwater resources problems.

Saltwater intrusion is also a serious problem in the large coastal cities of South and Southeast Asia, where groundwater abstraction is widespread but unregulated. The depression of groundwater tables because of overabstraction has caused saltwater intrusion as far inland as 10 kilometers in the coastal alluvial aquifer in Madras, and up to 5 kilometers in Manila. In Bangkok, groundwater overexploitation has caused the water level of the underlying aquifer to drop by 60 meters, resulting in problems with saltwater intrusion and land subsidence. The land surface has fallen by as much as 60 to 80 centimeters in the center of the city (British Geological Survey 1996:21).

Finally, although little information is available, changes in the water levels affect cave- and aquifer-dwelling species. Most of these species have not yet been studied and may have important ecological functions (McAllister et al. 1997:24).

Wetland Extent and Change

Wetlands are a key component of freshwater ecosystems. They include a variety of highly productive habitat types from flooded forests and floodplains to shallow lakes and marshes. Wetlands provide a wide array of goods and services, including flood control, nutrient cycling and retention, carbon storage, water filtering, water storage and aquifer recharge, shoreline protection and erosion control, and a range of food and material products, such as fish, shellfish, timber, and fiber. Wetlands also provide habitat for a large number of species, from waterfowl and fish to invertebrates and plants. In North America, for instance, 39 percent of plant species depend on wetlands (Myers 1997:129). Wetlands also have aesthetic and recreational values, although these are harder to quantify. They can include activities, such as birdwatching, hiking, fishing, and hunting.

Not only are wetlands highly productive and biologically rich, but a large part of the world's population lives in or near floodplain areas, where the soils are rich in nutrients and, therefore, very fertile. As a result of their potential as agricultural land (and also because they are feared as places that harbor disease), wetlands have undergone massive conversion around the world. Sometimes, this has come with considerable ecological and socioeconomic costs (*see Box 1 on the Okavango Delta*). In Africa, for example, many wetlands and floodplains are at risk of disappearing as a result of large-scale irrigation schemes and other water management activities (Barbier and Thompson 1998:434). When large irrigation or hydropower schemes are proposed, the downstream impacts often are not fully taken into account. Barbier and Thompson (1998) compared the gains upstream to the losses downstream of a large-scale irrigation scheme in the Hadejia-Jama'are River basin in northeastern Nigeria. This analysis concluded that even though the proposed irrigation scheme would increase food production upstream, the water management plan would not provide enough benefits to justify the significant losses to local communities downstream that would result from reduced agricultural, fishing, and fuelwood production (Barbier and Thompson 1998:439–440).

Another dramatic example of the unforeseen damage by large-scale water engineering works is the effects of a dam and flood embankments on the Waza-Logone floodplain in northern Cameroon (Ngantou and Braund 1999:19, 20). In the 1970s, the Waza-Logone floodplain provided habitat for tens of thousands of grazing mammals, birds, and fish, as well as sustenance for more than 100,000 sedentary and nomadic people. The population lived primarily by fishing and pastoralism — activities that depended heavily on the annual flood regime. The construction of a dam and the development of a rice irrigation scheme in 1979 resulted in reduced flooding over large parts of the plain, leading to the collapse of the fishery, a reduc-

Box 1

What Is at Stake and What Are the Trade-offs: Okavango Delta

One of the most remarkable cases in which a globally important wetland area is at risk from drainage and water extraction is the Okavango Delta in Botswana. The Okavango River flows from the highlands of Angola through Namibia and Botswana, where it forms the Okavango Delta, a vast 15,000-square kilometer alluvial fan on the otherwise parched sands of the Kalahari Desert (Rothert 1999:1). The desert oasis provides habitat for a large array of plants and animals that would not survive the arid environment, and it helps sustain almost 130,000 people of several ethnic groups living in over 30 communities (Rothert, personal communication, 2000). The people have depended on the Okavango's waters for generations, and with a growing tourist industry, which employs thousands of people and earns Botswana over US$250 million in revenues annually, they have every incentive to preserve and manage the Delta's wildlife and water resources (Rothert 1999:1).

The Okavango River is also one of the few large river systems in the world without notable human developments such as dams or water diversions. But over the past 20 years, there have been several proposals for large-scale engineering projects that would alter this fragile ecosystem and develop the region. The first large development plan for the Delta was the 1982 Southern Okavango Integrated Water Development Project. The Botswana government faced strong opposition to the project from local residents, and it eventually agreed to an outside assessment of the project that was conducted by IUCN-The World Conservation Union (IUCN). The IUCN report concluded that most of the major engineering works of the project would not meet their stated objectives of increasing food production and rural living standards. The Botswana government cancelled all plans to implement the project in 1992 over the protests of the bureaucracy in charge of water development (Howe 1994:25, 28).

Large development projects that will likely cause harm to the Delta ecosystem are still being proposed. In late 1996, the government of Namibia floated plans to build a 250-kilometer pipeline from the Okavango River to tap 20 million m^3 of the river's water annually for Namibia's capital city (Pottinger 1997:1). This project would be the first major diversion of the Okavango's waters, and it would have unknown but potentially negative effects on wildlife, rural livelihoods, and the Delta's growing tourist industry. A recent review of Namibia's water needs showed that groundwater sources, combined with existing water supplies, effective demand management, and artificial aquifer recharge, could provide the needed water to Namibia without the costly development of the pipeline (Rothert 1998:22). Yet pressure to build the pipeline arises every few years with the onset of a drought, a regular occurrence in this arid region. A short-term planning perspective means that the likelihood of diverting the Okavango River remains high (Rothert 1998:1, 2).

References

Howe, C.W. 1994. "The IUCN Review of the Southern Okavango Integrated Water Development Project." *Environment* 36 (1):25–28.

Pottinger, L. 1997. "Namibian Pipeline Project Heats Up." *World Rivers Review* 12 (1):2–3. Available on-line at: http:\\www.irn.org.

Rothert, S. 1998. "Okavango Pipeline Remains a Pipe-Dream for Now." *World Rivers Review* 13 (3):1–2. Available on-line at: http:\\www.irn.org.

Rothert, S. 1999. *Meeting Namibia's Water Needs While Sparing the Okavango.* Berkeley, California, U.S.A.: International Rivers Network.

Rothert, S. International Rivers Network. Personal communication, July 20, 2000.

tion in available grazing grounds, and a shortage of surface water during the dry season. As a consequence, there was a drastic decline in biodiversity as well as a massive migration of people and livestock out of the area. In 1988, the international community recognized the problems in the Waza-Logone floodplain and helped launch a program to rehabilitate the floodplain. The rehabilitation project is now in its third phase, and it already has boosted fish production and increased the availability of dry-season grazing (Ngantou and Braund 1999:19, 20).

GLOBAL EXTENT OF WETLANDS

A recent review of wetland resources concluded that reliable global estimates of wetlands extent could not be produced with available data (Finlayson and Davidson 1999:2). The review—commissioned by the Convention on Wetlands, also known as the Ramsar Convention, and carried out by Wetlands International and the Environmental Research Institute of the Supervising Scientist in Australia – stated that regional data for Oceania, Asia, Africa, eastern Europe, and the neotropics allow just a cursory assessment of wetland extent and location. Only North America and western Europe have published more robust estimates of wetland extent. In addition, some good examples of wetland inventory processes exist. The MedWet program, for example, is coordinating wetlands activities throughout the Mediterranean (Finlayson and Davidson 1999:6).

Wetland inventories have been incomplete and are difficult to undertake for a number of reasons:

- Definitions. The definition of a wetland varies with research purpose and an organization's mandate. For example, studies interested in modeling trace gases have concentrated on inland wetlands (defined as "areas with a water table at or near the soil surface for a significant part of the growing season, and which are covered by active vegetation during the period of water saturation," IGBP 1998:7). The Ramsar Convention uses a very broad definition, covering both inland and marine wetlands. The latter includes reefs and seagrass beds as specific habitat types because the definition delineates marine areas up to a water depth of six meters.
- Scope. Data from national inventories are often incomplete and difficult to compare because some concentrate on specific habitat types, such as wetlands of importance to migratory birds, whereas others include artificial wetlands, such as rice paddies.
- Limitations of maps. Wetland inventories sometimes use existing maps to estimate wetland extent, such as Operational Navigation Charts for global analyses and large-scale topographic maps for national assessments. Wetland extent then becomes a function of scale and the cartographic convention of the mapmaker. For example, navigation charts will depict only wetlands that are visible by pilots. In addition, the rules for placing wetlands symbols on a map usually vary with institutions.
- Boundaries. Researchers have problems defining boundaries and separating individual wetlands from wetland complexes. Some boundaries are difficult to determine because of limited access and seasonal changes in water availability.
- Limitations of remote sensing products. The wide range in the sizes and types of wetlands and the problem of combining hydrologic and vegetation characteristics to define wetlands make it difficult to produce a global, economical, and high-resolution data set with existing sensors.

Considering these limitations, Map 4 is currently the best regional approximation of wetlands in Africa. The World Conservation Monitoring Centre (WCMC) and IUCN— estimated the location and extent of the wetlands. A group of experts delineated wetlands boundaries by generalizing information on inundated areas, rivers, lakes, and topography from the 1:1 million Operational Navigation Charts (WRI 1995:18). This map provides more detail than other global, coarse resolution data sets that use potential vegetation, soils, and terrain to delineate wetlands—for example, data produced to estimate methane emissions (Matthews and Fung 1987:61–86) or the International Satellite Land Surface Climatology Project (ISLCP 1987–1988). It also provides more detail on wetlands than the most recent land cover characterization map by the International Geosphere-Biosphere Programme (IGBP) of the International Council for Science, which mostly shows coastal wetlands. However, because of its scale, it underestimates wetlands areas in valley bottoms, such as *dambos* ("valley meadowlands" in southern Africa), which are important for agricultural production, food security, and habitat (Chenje and Johnson 1996:52–54).

Map 4 also shows a proxy variable, location of dams, to indicate the potential of changed hydrological regimes. Only a few of the wetlands shown on Map 4 have been designated as important wetlands under the Ramsar Convention (Ramsar Convention Website: http://ramsar.org/key_sitelist.htm). Of these, 6 have been listed in the Montreux Record, a register of wetland sites on the List of Wetlands of International Importance that includes places where changes in ecological character have occurred, are occurring, or are likely to occur as a result of technological developments, pollution, or other human interference (Ramsar Convention Website: http://ramsar.org/key_montreux_record.htm).

HOW HAS WETLAND AREA CHANGED?

Modifying wetlands has been a major focus of many development and river regulation plans for decades. Wetlands have been either completely converted to other land uses (often by building drainage ditches and filling in swamps) or their functions have been altered gradually by changing hydrologic regimes and introducing agricultural crops and livestock. Data on actual numbers of wetlands converted or modified are not available globally. However, Myers (1997:129) estimated that half of the wetlands of the world were lost in the 20th century. A 1992 review of 344 Ramsar sites showed that 84 percent were either threatened or experiencing ecological changes, mostly from drainage for agriculture and urban development, pollution, and siltation (Dugan and Jones 1993:35–38).

Certain regions and countries have undertaken a more concise effort to track wetland loss. For example, the U.S. Fish and Wildlife Service tracks the status of U.S. wetlands and the amount of wetlands lost or gained over time through its National Wetlands Inventory. Its first assessment examined the extent of wetland losses over the first 200 years of the country's history. Total wetland losses in the lower 48 states were estimated to be 53 percent from the 1780s to the 1980s, with 22 states believed to have lost 50 percent or more of their wetlands (*see Map 5a*). Only Alaska, Hawaii, and New Hampshire lost less than 20 percent (Dahl 1990:5). Most of the losses were to agriculture.

Wetland losses in the lower 48 United States were also quantified and statistically compared for the 1970s and 1980s (Dahl and Johnson 1991:1–23). The study estimated the loss of wetlands in the lower 48 states during this period to be greater than 2 percent, totaling 1.05 million hectares. The total recorded change in wetlands area dropped from 42.9 million hectares in the mid-1970s to 41.8 million hectares in the mid-1980s (Dahl and Johnson 1991:1).

The study showed that 1.34 million hectares of inland freshwater wetlands were lost during this period, while 320,680 hectares of freshwater ponds were gained. Of the lost freshwater wetlands, 54 percent were converted to agriculture, 41 percent were filled but not yet converted to an identifiable land use category, and 5 percent were converted to urban uses (Dahl and Johnson 1991:11, 12). The largest drop of any wetland category was for forested freshwater wetlands, which declined by 6.2 percent over the ten-year period. Much of this loss was in the southeastern states, each of which lost significantly more than 40,000 hectares of forested freshwater wetlands (Dahl and Johnson 1991:11, 12).

The Natural Resources Inventory (NRI), a division of the U.S. Department of Agriculture, carried out an additional study of wetland loss on nonfederal lands from 1982–87 in the United States. During this period an average of 72,600 hectares were lost each year (Brady and Flather 1994:693). Although this analysis was limited to the five-year period, it provides some indication of regional trends in wetland losses and gains during the 1982–92 period that Map 5b illustrates.

The combined NRI and Fish and Wildlife Service data give the best estimates of wetland losses over time in the United States. These data show that the rate of wetland losses has declined substantially in the last 50 years. The estimate of average annual loss from the time of settlement to 1954 is 329,000 to 359,000 hectares. This dropped to 185,000 hectares per year from 1954 to 1974. From 1974 to 1982 the rate dropped again, to 117,000 hectares yearly, and then to 32,000 hectares yearly from 1982 to 1992 (Heimlich et al. 1998:20).

In Europe, particularly in the Mediterranean basin, wetland loss is even more severe. Estimates show, for example, that Spain has lost more than 60 percent of all inland freshwater wetlands since 1970 (EEA 1995:207); Lithuania has lost 70 percent of its wetlands in the last 30 years (EEA 1999:291); and the open plains of the southwestern part of Sweden have lost 67 percent of their wetlands and ponds to drainage in the last 50 years (EEA 1995:185). Overall, draining and conversion to agriculture alone has reduced wetlands area in Europe by some 60 percent (EEA 1998:291).

Wetland loss data for other regions of the world, as well as data on more gradual modifications of wetlands hydrological regime are even harder to obtain. For example, because wetland ecosystems depend on a saturated water table and are generally very sensitive to changes in water levels, groundwater overabstraction can cause substantial damage to wetlands by causing the underlying water table to drop. If water withdrawals are large enough, wetlands can be permanently destroyed (EEA 1995:66). In 1995, the European Environment Agency estimated that around 25 percent of the most important wetlands in Europe were threatened by groundwater overexploitation (EEA 1995:67). A more recent study shows that six wetlands designated as Ramsar sites in Denmark, four sites in Hungary, and one in the United Kingdom were reported as being in danger because of groundwater overexploitation (Scheidleder et al. 1999:96). An additional 46 Ramsar sites from 7 European countries were reported as being under threat for other reasons (Scheidleder et al. 1999:96). The information gathered from these countries, however, is very incomplete. Only 14 countries out of 37 responded to the questionnaire, and the threat status for wetlands was provided for only 50 percent of the identified Ramsar sites. Therefore, these data do not reflect the actual extent of the threat to wetlands on the continent (Scheidleder et al. 1999:95, 96).

OPPORTUNITIES FOR FUTURE WETLANDS MONITORING USING REMOTE SENSING

The IGBP has identified the production of a global baseline data set of wetlands as a high priority (IGBP 1998). New remote sensing techniques and improved capabilities to manage complex global data sets make such an endeavor more feasible (IGBP 1998:17). Radar, which can sense flooding underneath vegetation and can penetrate cloud cover, is probably the best alternative for developing a global wetland database (IGBP 1998:27). Because of its high resolution and sensitivity to water, radar data reveal a much finer wetland texture, particularly in areas of flooded forests, than other remotely sensed data. Consequently radar data show much larger wetland areas than existing estimates, which is consistent with the conclusions of the IGBP assessment that the global extent of wetlands has been significantly underestimated (Darras et al. 1999:39). Seasonally flooded wetlands provide highly productive croplands and grazing, essential for food security and income generation, as well as providing important habitat for migratory birds and other terrestrial and aquatic species. Information on the extent, location, and change in wetland areas is, thus, crucial to assessing the condition of freshwater ecosystems.

The launch of Landsat 7 in April 1999 opened a new era in civilian remote sensing, thanks to a higher quality sensor, a comprehensive data acquisition strategy, and a liberal data policy. The sensor includes a 15-meter resolution panchromatic band, in addition to five 30-meter resolution optical and infrared bands, which can be used to resolve more detailed landcover features. Landsat 7 also has much greater onboard storage capacity than previous Landsat generations which, coupled with an acquisition strategy based on cloud climatology, is intended to provide complete global coverage four times a year.

Finally, in an attempt to maximize use of these data, the price per image has been reduced from US$4,200 to US$600. Higher quality data, available on a regular basis and at modest cost, should allow complete and accurate mapping of the extent of both seasonal and permanent wetlands (USGS Landsat 7 Website: http://landsat7.usgs.gov/).

Watershed Modification

Freshwater systems are influenced not only by modifying rivers, lakes, and wetlands directly, but also by changing land-use patterns in the whole watershed. The pattern and extent of cities, roads, agricultural land, and natural areas within a watershed influences infiltration properties, transpiration rates, and runoff patterns, which in turn impact water quantity and quality. For example, expanding impervious areas increases the volume and rate of runoff of receiving streams and impacts the water quality and biodiversity of freshwater systems (Jones et al. 1997:51).

At the global level, we used two indicators representing agricultural and urban areas to show the extent to which watersheds have been modified in ways likely to affect the condition of freshwater ecosystems. First, we examined the distribution of watersheds containing intensive agriculture because watersheds with intensive agricultural development are likely to experience water quality degradation from pesticide and nutrient runoffs and increased sediment loads (*see Map 6*). Agricultural areas were extracted from the IGBP version of a global one-kilometer resolution land-cover characteristics database (GLCCD 1998). These cropland areas exclude those with more balanced mosaics of cropland and natural vegetation. Second, we examined the distribution of urban areas as judged by satellite images of nighttime lights for 1994–95. Because more urbanized watersheds tend to have greater impervious areas as well as higher quantities of urban and industrial pollution, this map also shows greater pressure on freshwater systems (*see Map 7*). Urban areas came from a five-kilometer resolution stable-lights database (NOAA-NGDC 1998). The stable-lights database measures emitted nighttime visible and near-infrared radiation from cities, towns, and industrial sites. The data better represent urban areas with highly developed economies indicated by extensive electricity networks, street lighting, and industrial activities, such as refineries. They underestimate urban areas within countries with less developed economies.

These two indicators show the degree of human modification of original natural vegetation in a watershed. They are also proxy indicators for water quality impacts, indicating the potential of pesticide and nitrogen runoffs, increased sediment loads, and increased potential for urban and industrial pollutants. All indicators were aggregated by large river basins to produce global maps.

Maps 6 and 7 show contrasting patterns of modified land use. Map 6 shows that intensively cropped land is concentrated in five areas: Europe, India, eastern China, Southeast Asia, and the midwestern United States, with smaller concentrations in Argentina, Australia, and Central America. Africa is striking in its lack of intensively cropped land, with the exception of small patches along the Mediterranean coast and in South Africa. This reflects the minimal use of chemical inputs and the low level of agricultural productivity in most African countries.

Map 7 shows a much more concentrated pattern of urban and industrial development. Highly urbanized watersheds are concentrated along the east coast of the United States, Western Europe, and Japan, with lesser concentrations in coastal China, India, Central America, most of the United States, Western Europe, and the Persian Gulf.

These maps show the average amount of intensively cultivated land or urban and industrial area for each large river basin. However, they do not allow for the assessment of land use within watersheds and, therefore, fail to show potentially important within-basin differences. We, therefore, used a higher resolution river basin database to calculate the concentration of intensively cultivated land within individual subbasins (EDC 1999). Map 8a shows the results for Europe (west of the Ural Mountains) and the Middle East; and Map 8b shows the results for insular Southeast Asia. There are 4,033 subbasins in Europe and 4,077 in insular Southeast Asia.

In Europe, the most intensively cultivated land area forms an arc extending from northern France to the Ukraine. Crop intensity is higher in basins in northern France, the Netherlands, and southern England, and in the subbasins of the Oder, Vistula, Dnieper, and Don rivers in Eastern Europe. There are also intensively cropped areas in parts of the Danube basin and subbasins close to the Black Sea, particularly around the Sea of Azov.

Insular Southeast Asia presents a gradient of agricultural use from more populated areas, such as Java and the Philippines, where intensively cultivated areas are predominant in most basins, to a low level of cropland area in Borneo, and the Celebes.

Modification of Freshwater Systems: Information Status and Needs

The previous section demonstrates the influence of humans on the hydrological cycle. This influence is global and the degree of modification is significant. As the river fragmentation and sediment retention indicators show, there are very few large, free-flowing river systems in the world. While dam construction has greatly slowed in most developed countries, it continues apace in China, Japan, India, Iran, Turkey, and several countries in South America. Our understanding of the environmental and social impacts of such projects is constrained by the limited availability and patchy quality of information. Several information gaps need to be filled to get a more accurate assessment of the condition of freshwater systems around the world.

One of the key databases needed to assess the condition of freshwater systems is a complete global database on dams. The information used to develop the indicators in this analysis was restricted to large dams, and it came mostly from ICOLD's World Register of Dams (ICOLD 1998). The registry contains information only for dams that are 15 meters in height or greater, except for China, Japan, India, Spain, and the United States where entries are only for dams 30 meters in height or greater (ICOLD 1998:7). The largest data gaps are for Russia, which reports only hydropower dams, and China, where the majority of the world's large dams have been built and for which information is exceedingly difficult to acquire.

The dams used in this study represent only a portion of the global water impoundment and are missing basic attributes about dam operations and biophysical characteristics of the reservoirs (Vörösmarty et al. 1997a:218). For example, data on downstream discharges are lacking for many reservoirs, however, these data are needed to assess more fully the interannual variations in river flow brought about by engineering. Current research is, therefore, limited to statements based on mean annual conditions, which miss important dry season variations (Vörösmarty and Sahagian 2000:13). This information is sometimes available for individual reservoirs, but there is a lack of basic documentation at continental or global scales.

The existing global register of dams is also limited by a lack of data about dam locations. The ICOLD database, for example, provides the river and closest town for each entry, but no latitudinal and longitudinal coordinates, which makes the process of geo-referencing dams very tedious and adds to location errors. The provision of latitude and longitude for each dam would highly improve our ability to locate these structures within the correct hydrological unit and assess their impact on the freshwater systems (Vörösmarty and Sahagian 2000:14).

Another key data set needed to assess freshwater ecosystem conditions is complete global information on wetland distribution and change. There is sufficient demand by international conventions (such as Ramsar and the Convention on Biological Diversity) and scientists working on global environmental issues to improve global data on wetlands. The recent review of wetland resources, commissioned by the Ramsar Convention (Finlayson and Davidson 1999), made two important recommendations:

- Focus first on producing basic data on the location and size of wetlands, especially for Asia, Africa, South America, the Pacific Islands, and Australia, using remotely sensed data and surveys. The data need to include major biophysical features, including information on the variation of wetland areas and the timing of floods.
- In a second step, compile data on wetland threats, land tenure, management, and uses and initiate studies on wetland benefits and values.

These recommendations were echoed at a March 1999 planning meeting of the IGBP, which emphasized that the production of a global wetland inventory was a high priority (Darras et al. 1999:3).

Finally, global data on the exploitation and condition of groundwater resources are also urgently needed. Many countries that depend on groundwater resources have inventoried the location of aquifers and their use. On the continental and global levels, however, such information is inconsistent and not readily available. Data on groundwater quality is extremely limited and available only for some countries. Countries that share major aquifers should coordinate information on the resource. Information on groundwater exploitation should also be collected in coordination with data collection efforts on the effects of these extractions on other regional water resources, such as wetlands, lagoons, and river basins.

WATER QUANTITY

Overview

Water, used by households, agriculture, and industry, is clearly the most important good provided by freshwater systems. Humans now withdraw about 4,000 km^3 of water a year, or about 20 percent of the world's rivers' base flow (the dry-weather flow or the amount of available water in rivers most of the time) (Shiklomanov 1997:14, 69). Between 1900 and 1995, withdrawals increased by a factor of more than six, which is greater than twice the rate of population growth (WMO 1997:9).

Scientists estimate that the average amount of global runoff (the amount of water that is available for human use after evaporation and infiltration takes place) is between 39,500 km^3 and 42,700 km^3 a year (Fekete et al. 1999:31; Shiklomanov 1997:13). However, not all of this water is available to humans. Much of the runoff occurs in flood events or is inaccessible to people because of its remote location. In addition, part of the runoff needs to remain in waterways so that aquatic ecosystems continue to function. In fact, only around 9,000 km^3 is readily accessible to humans, and an additional 3,500 km^3 is stored in reservoirs (WMO 1997:7).

In any case, such global averages fail to portray the details of the world's water situation. Water supplies are distributed unevenly around the globe, with some areas containing abundant water and others a much more limited supply. For example, arid and semiarid regions receive only 2 percent of the world's runoff, even though they occupy roughly 40 percent of the terrestrial area (WMO 1997:7). In water basins with high water demand relative to the available runoff, water scarcity is a growing problem.

According to the U.N. Comprehensive Assessment of the Freshwater Resources of the World, there are close to 460 million people in the world currently suffering from serious water shortages and an additional 25 percent of the world's population may become water-stressed if current consumption levels continue (WMO 1997:9). Many experts, governments, and international organizations around the world are predicting that water availability will be one of the major challenges facing human society in the 21st century and that the lack of water will be one of the key factors limiting development (WMO 1997:5).

Water is not only becoming scarce in many regions of the world because of increased demand by industries and municipalities, but with an expected population reaching about 7.8

billion by 2025— the U.N medium projection (UNPD 1999:2) —food production will have to increase, meaning more water resources will be needed for irrigation.

At present, irrigated agriculture accounts for 40 percent of global food production even though it represents just 17 percent of global cropland (WMO 1997:9). Agriculture is society's major user of water, withdrawing 70 percent of all water (WMO 1997:8). Most irrigation systems are relatively inefficient. Global estimates of irrigation efficiency are around 40 percent (Postel 1993:56; Seckler et al. 1998:25). Even though irrigation for food production is an increasingly important service, it is not analyzed in detail in this report because it has been examined separately under the agroecosystem component of PAGE (Wood et al. 2000).

At the same time that water demand is increasing, pollution from industry, urban centers, and agricultural runoff is limiting the amount of water available for domestic use and food production. In developing countries, an estimated 90 percent of wastewater is discharged directly to rivers and streams without any waste processing treatment (WMO 1997:11). In many parts of the world, rivers and lakes have been so polluted that their water is unfit even for industrial uses (WMO 1997:11). Threats of water quality degradation are most severe in areas where water is scarce because the dilution effect is inversely related to the amount of water in circulation. The loss of ecological services is considerable, but difficult to assess.

The Aral Sea represents one of the most extreme cases in environmental degradation of an aquatic system. Large-scale upstream diversions of the Amu Darya and Syr Darya river flow for irrigation of about seven million hectares of land, has reduced inflow to the extent that the volume of water in the Aral basin has been reduced by 75 percent since 1960 (UNESCO 2000:35; Postel 1999:94). This loss of water, together with excessive chemicals from agricultural runoff, has caused a collapse in the Aral Sea fishing industry, a loss of biodiversity and wildlife habitat, particularly the rich wetlands and deltas, and an increase in human pulmonary and other diseases in the area resulting from the high toxicity of the salt concentrations in the exposed seabed (Postel 1999:94; WMO 1997:10).

Condition Indicators of Water Quantity

Given humanity's dependence on water, one would expect that an assessment of the capacity of freshwater systems to provide this basic good would be easy to carry out. However, researchers looking into freshwater ecosystems at the global level face two major data gaps: water use and water supply at the watershed level. For some countries, reliable information on water supply and use is not even available at a national level.

To get a better understanding of the balance of water demand and supply and to better estimate the dimensions of the global water problem, the PAGE study developed two indicators that measure the capacity of freshwater systems to provide water for human consumption. These indicators are the following:

- Per capita water supply by river basin;
- Dry season flow by river basin.

PER CAPITA WATER SUPPLY BY RIVER BASIN

To calculate water supply, the PAGE study undertook a new analysis of water scarcity using a somewhat different methodology than the 1997 U.N. assessment mentioned previously. Because country-level estimates of water availability may hide significant within-basin differences, the PAGE study calculated water supply for individual river basins, rather than at a national or state level, with the object of identifying those areas where water supply per person fell below 1,700 m^3/year.

Water experts define areas where per capita water supply drops below 1,700 m^3/year as experiencing "water stress"—a situation in which disruptive water shortages can frequently occur (Falkenmark and Widstrand 1992:1–33; Hinrichsen et al. 1998:4). In areas where annual water supplies drop below 1,000 m^3 per person per year, the consequences can be more severe and lead to problems with food production and economic development unless the region is wealthy enough to apply new technologies for water use, conservation, or reuse.

According to the PAGE analysis, some 41 percent of the world's population, or 2.3 billion people, live in river basins under water stress, with per capita water supply below 1,700 m^3/year (*see Map 9 and Table 3*). Of these, some 1.7 billion people reside in highly stressed river basins where water supply falls below 1,000 m^3/year. By 2025, the PAGE analysis projects that, assuming current consumption patterns continue, at least 3.5 billion people— or 48 percent of the world's projected population —will live in water-stressed river basins (*see Map 10 and Table 3*). Of these, 2.4 billion will live under high water stress conditions. This per capita water supply calculation, however, does not take into account the coping capabilities of different countries to deal with water shortages. For example, high-income countries that are water scarce may be able to cope to some degree with water shortages by investing in desalination or reclaimed wastewater. The study also discounts the use of fossil water sources because such use is unsustainable in the long term.

The 2025 estimates are considered conservative because they are based on the United Nations' low-range projections for population growth, which has population peaking at 7.2 billion in 2025 (UNPD 1999:3). In addition, a slight mismatch between the water runoff and population data sets leaves 4 percent of the global population unaccounted for in this analysis.

Map 9 was developed by combining a global population database for 1995 that uses census data for over 120,000 admin-

Table 3

Global Annual Renewable Water Supply Per Person in 1995 and Projections for 2025

Water Supply *(m^3/person/year)*	1995 Population *(millions)*	1995 Percent of Total	2025 Population *(millions)*	2025 Percent of Total
<500	1,077	19.0	1,783	24.5
500–1,000	587	10.4	624	8.6
1,000–1,700	669	11.8	1,077	14.8
>1,700	3,091	54.6	3,494	48.0
Unallocated	241	4.2	296	4.0
Total	5,665	100	7,274	100

Source: WRI.

istrative units (CIESIN et al. 2000) and a global runoff database developed by the University of New Hampshire and the WMO/Global Runoff Data Centre (Fekete et al. 1999). The runoff database combines observed discharge data from monitoring stations with a water balance model driven by climate variables such as temperature, precipitation, land cover, and soil information. For those regions where discharged data were available, the modeled runoff was adjusted to match the observed values; for regions with no observed data the modeled estimates of runoff were used (Fekete et al. 1999).

This runoff model provides a slightly lower estimate of global runoff than previous analyses—39,500 km^3/year (Fekete et al. 1999:31) instead of 42,700 km^3/year (Shiklomanov 1997:13) or 47,000 km^3/year (Seckler et al. 1998:3). Based on these data, the percentage of the population that is estimated to be living in conditions of water stress in 1995 is higher than previous estimates. This is due in part to the slightly lower estimate of global runoff and a more current subnational population data set for 1995 (CIESIN et al. 2000). Finally by using a river basin approach instead of a national level analysis, many within-country differences in water supply can be highlighted.

The results of this analysis also show that of those basins where the projected population is expected to be higher than 10 million by 2025, 6 basins will go from having more than 1,700 m^3 to less than 1,700 m^3 of water per capita per year. These basins are the Volta, Farah, Nile, Tigris and Euphrates, Narmada, and the Colorado River basin in the United States (*see Maps 9 and 10*). Another 29 basins will descend further into scarcity by 2025, including the Jubba, Godavari, Indus, Tapti, Syr Darya, Orange, Limpopo, Huang He, Seine, Balsas, and the Rio Grande (*see Maps 9 and 10*).

Other water availability projections suggest similar trends of increasing scarcity. For example, the WaterGAP model developed by the University of Kassel in Germany estimates that under a "business as usual" scenario four billion people in 2025 will live in areas experiencing severe water stress (Alcamo et al. 2000:3). The business as usual scenario assumes that "current trends in population, economy, technology, and human behavior continue up to 2025" (Alcamo et al. 2000:25). According to the model's results, increased scarcity will be especially marked in South and Southeast Asia and southern and western Africa.

Although the estimates of percent population experiencing scarcity are similar, these results are not directly comparable with the PAGE analysis because the modeling approaches are very different. The PAGE model assumes constant water supply, with human population growth as the main cause of rising water demands. It also uses benchmarks of per capita available water to determine which watersheds will experience shortages (Falkenmark and Lindh 1993). These estimates of per capita water demand are conservative; they assume that as much as 30 to 50 percent of the freshwater runoff can be mobilized for human consumption. These benchmarks also base their agricultural water demand on a "nutritious, low-meat diet" (FAO in Postel 1997), and twothirds of crop water being supplied by rainwater.

WaterGAP uses a very different approach that integrates climate and land cover changes into future water supply, and also factors economic growth and technological advances, as well as population growth, into water demand projections. WaterGAP uses a higher population projection for 2025 (8 billion people) and sets population and income level as the main drivers behind increases in domestic water demand. However, the largest increase in water consumption in its model comes from industry.

Global concerns about water scarcity include not only surface water sources but groundwater sources as well. More than 1 billion people in Asian cities and 150 million in Latin American cities rely on groundwater (Foster et al. 1998:xi). In addition, although there are no complete figures on groundwater use by the rural population, many countries are increasingly dependent on this resource for both domestic and agricultural use (Foster et al. 2000:1). Currently humans withdraw approximately 600–700 km^3 per year—about 20 percent of global water withdrawals (Shiklomanov 1997:53–54). Some of this water is fossil water that comes from deep sources that are isolated from the normal runoff cycle, but much groundwater comes from shallower aquifers that draw from the same global runoff that feeds freshwater ecosystems. Indeed, overdrafting of groundwater sources can rob streams and rivers of a significant fraction of their flow. In the same way, pollution of aquifers by nitrates, pesticides, and industrial chemicals often affects water quality in adjacent freshwater ecosystems. Although overdrafting and contamination of groundwater aquifers are known to be widespread and growing problems (UNEP 1996:4–5), comprehensive data on groundwater resources and pollution trends are not available at the global level.

DRY SEASON FLOW BY RIVER BASIN

These basinwide estimates mask the effect of seasonal runoff patterns, which reduce that amount of water that is available for human use. In many parts of the tropics, rainfall is highly concentrated in time. For example, it has been estimated that all of India's rainfall falls in 100 hours during the monsoon months (Seckler, personal communication, 1999). Most runs off into the sea before it can be captured for human use.

In almost every continent, river modification has affected the natural flow of the rivers to a point where, during the dry season, the outflow to the sea is nonexistent. For example, such rivers as the Colorado, Huang-He, Ganges, Nile, Syr Darya, and Amu Darya, all run dry at the mouth during the dry season (Postel 1995:10). The Amu Darya and the Syr Darya used to contribute 55 billion cubic meters of water annually to the Aral Sea prior to 1960, but this volume has been reduced to an average of 7 billion m^3, or 6 percent of the total annual flow for the period 1981–90 (Postel 1995:14, 15). Most of the water is diverted for irrigation (Postel 1995:14).

To assess runoff seasonally, and its implications for water availability, base flow as a percent of total flow per person was calculated for every river basin. Base flow, the volume of runoff available during the dry season (Foster et al. 1998:10), is defined here hydrologically as the four consecutive months with the lowest cumulative runoff. Map 11 highlights basins that are either water stressed (less than 1,700m^3/year per person) or have just adequate supplies of water (between 1,700 and 4,000 m^3/year per person) with a pronounced dry season. Basins with a pronounced dry season are those where less than 2 percent of the total annual runoff occurs in the 4 driest months of the year. Twenty-seven of these basins had more than ten million people in 1995 (*outlined in black in Map 11*). They include the Balsas and Grande de Santiago basins in Mexico, the Limpopo in Southern Africa, the Hai Ho and Hong in China, the Chao Phraya in Southeast Asia, and the Brahmani, Damodar, Godavari, Krishna, Mahi, Narmada, Ponnaiyar, Rabarmarti, and Tapti in India. Here, low dry season flows have exacerbated water supply and quality problems.

Capacity of Freshwater Systems to Provide Water

Humans withdraw about one fifth of the normal (nonflood) flow of the world's rivers, but in river basins in arid or populous regions the proportion can be much higher. This has implications for the species living in or dependent on these systems, as well as for future human water supplies. Currently, more than 40 percent of the world's population lives in water-scarce river basins. With growing populations, water scarcity is projected to increase significantly in the next decades, affecting half of the world's people by 2025. Widespread depletion and pollution of groundwater sources, which account for about 20 percent of global water withdrawals, also is a growing problem for freshwater ecosystems because groundwater aquifers are often linked to surface water sources.

Water Quantity Information Status and Needs

The single greatest barrier to better analyses of how the hydrological cycle is being impacted by engineering works and land-use change is the poor quality of the hydrological data. The reliability and availability of hydrological data has deteriorated sharply since the mid-1980s when international support of national monitoring programs was reduced, particularly from U.N. agencies. The number of functioning hydrological stations has fallen significantly since 1985 (Fekete et al. 1999:8). In many parts of the developing world we know less about hydrological conditions than we did 20 years ago. As a result, major projects are being designed and in some cases implemented without the basic hydrological data needed to assess the financial—let alone environmental—impacts of these projects.

In addition, currently available statistics on water withdrawal and consumption are fraught with uncertainty because of the highly decentralized nature of water use. Our knowledge of the extent and nature of irrigated land at the continental and global levels is poor, for example, even though the agriculture sector is responsible for 93 percent of water consumption and 70 percent of withdrawals (WMO 1997:8). The irrigated area database developed by the University of Kassel and used in the PAGE agroecosystems report (Wood et al. 2000) is the only globally complete and consistent coverage that estimates area "equipped" for irrigation. However, the coarse resolution (50 kilometers) and lack of information about crop types, operations of irrigation systems, and other variables precluded its use for more detailed analyses. Even when higher resolution data are used, huge differences have been observed between remotely sensed estimates and field-based estimates. These differences are partly attributable to different definitions of irrigated land (for example, the area equipped for irrigation, the area actually irrigated, or the irrigated area multiplied by the number of crops) and underreporting by farmers (Frolking et al. 1999: 407–416).

As mentioned in previous sections, comprehensive statistics on supply and use of groundwater resources are also lacking for domestic and urban use, as well as irrigation (Foster et al. 1998:xi; Foster et al. 2000:2). In order to improve our ability to monitor the condition of freshwater systems to provide water for humans and ecosystems, better statistics on water availability and use are urgently needed—preferably at the watershed level.

Water Quality

Overview

The definition of water quality is not objective, but is socially defined depending on the desired use of water. Different uses require different standards of water quality. Water used for hydropower generation, industrial purposes, and transportation does not require high standards of purity. Such uses as recreation, fishing, drinking, and habitat for aquatic organisms rely on higher levels of water quality (UN/ECE 1995:5, 6). For that reason, water quality should be taken to mean the "physical, chemical, and biological characteristics of water necessary to sustain desired water uses" (UN/ECE 1995:5).

Monitoring the quality of water is important because clean water is necessary for human health and the integrity of aquatic ecosystems. Ecosystems filter and cleanse water. For instance, wetlands provide a very important service because they filter water by intercepting surface runoff, trapping sediments, and removing nitrogen and minerals from the water. This water filtering service has been estimated to be worth US$3 million a year for just a 5.5-kilometer stretch of the Alchovy River in the State of Georgia in the United States (Lerner and Poole 1999:41). This ability to filter and purify water, however, is being impaired by pollution and habitat degradation in many rivers, lakes, and estuaries around the world. As much as 3.3 billion people still lack access to adequate sanitation and more than a billion people lack access to safe drinking water, leading to millions of deaths and illnesses each year, mostly in the developing world (Cosgrove and Rijsberman 2000:9 and WHO 1996).

Most water quality monitoring was originally conducted at known sources of pollution, but this approach failed to detect the many diffuse nonpoint sources of water pollution. To overcome this deficiency, water quality monitoring has evolved along two different lines. One is a river basin approach. In the United States, this approach is used by the Environmental Protection Agency (EPA) in its Index of Watershed Indicators (IWI), and by the U.S. Geological Survey (USGS) with the National Water-Quality Assessment (NAWQA) Program. The IWI is briefly described in Box 2. The NAWQA program provides data on nutrients, pesticides, and volatile organic compounds for surface and groundwater in 60 important river basins and aquifers across the country. These programs generally examine only the physical and chemical qualities and quantities of water. One potential problem with river basin-level approaches is that such basins often cross state and national boundaries, necessitating the

Box 2

Index of Watershed Indicators from the U.S. Environmental Protection Agency

To better communicate water pollution problems and address water quality issues in the United States, the Environmental Protection Agency (EPA) has developed an Index of Watershed Indicators (IWI) for 2,262 watersheds. It integrates 15 indicators of watershed condition and vulnerability. The condition indicators are based on a range of variables, including the number of fish consumption alerts by watershed, data on contaminated sediments from sampling stations, recent and historical loss of wetlands, and water quality reports measured against state and tribal water quality standards designated for that particular water body.

In addition, the EPA uses another eight indicators to describe watershed vulnerability. These vulnerability indicators represent pressures on freshwater systems that are linked to water degradation and habitat quality in watersheds. They include the number of aquatic or wetland species at risk, the discharged loads of pollutants, and the potential impact of urban and agricultural runoff.

The EPA uses different weighting schemes to aggregate indicators of watershed condition and vulnerability into the IWI, assigning scores for each watershed assessed. Based on these scores, watersheds are classified into seven categories:

1. Watersheds with better water quality and lower vulnerability to stressors, such as pollutant loadings;
2. Watersheds with better water quality and higher vulnerability to stressors;
3. Watersheds with less serious water quality problems and lower vulnerability to stressors;
4. Watersheds with less serious water quality problems and higher vulnerability to stressors;
5. Watersheds with more serious water quality problems and lower vulnerability to stressors;
6. Watersheds with more serious water quality problems and higher vulnerability to stressors;
7. Watersheds with insufficient data to be assessed for water quality and vulnerability.

There has to be a minimum of 10 of the indicators in order for a watershed to get an "IWI score." Otherwise the watershed is characterized as having "insufficient data."

The following table presents the results for 1999:

IWI	Number of Watersheds
Better Water Quality–Low Vulnerability	310
Better Water Quality–High Vulnerability	29
Less Serious Water Quality Problems–Low Vulnerability	736
Less Serious Water Quality Problems–High Vulnerability	58
More Serious Water Quality Problems–Low Vulnerability	496
More Serious Water Quality Problems–High Vulnerability	38
Insufficient Data	595

Overall 32 percent of those watersheds with sufficient data to be scored are classified as having serious water quality problems and 48 percent are classified as having less serious water quality problems. It is also important to note that even in the United States, where data collection on water quality is more systematic than in many other countries, regulators lack the information to assess 26 percent of the watersheds selected for this index.

Source: EPA 1999.

need for a high level of governmental or intergovernmental coordination of monitoring activities.

Another approach to water quality monitoring is an integration of chemical and biological parameters to measure condition. Whereas biological measures of water quality do not eliminate the need for chemical water monitoring, they are often less expensive than chemical analyses and can provide a useful index of potential chemical pollution problems (UN/ECE 1995:9, 10). Biological monitoring programs are used to measure water quality standards in the United States, the United Kingdom, and Australia, and have been applied in a number of other countries around the world.

This section will examine different measures of surface and groundwater quality in use around the world, and will briefly examine trends in the water quality of rivers and streams. It will focus on Europe and the United States because trend data for other regions of the world are not as readily available (Shiklomanov 1997:27).

The quality of surface waters in industrialized countries has generally been improving with respect to some pollutants over the last 20 years, but new chemicals are increasingly becoming a problem. For example, in most developed countries waste treatment plants have considerably reduced fecal contamination of surface waters. However, in developing countries sewage treatment is still not the norm, with 90 percent being discharged directly into rivers, lakes, and coastal areas without any treatment (WRI 1996:21). Consequently, water-related diseases, such as cholera, amoebic dysentery, schistosomiasis, malaria, and trypanosomiasis among others, claim 5 million lives annually worldwide and cause illness in perhaps half of the population of the developing world each year (WHO 1996). New pollution problems from agricultural and industrial sources have emerged in both industrialized and developing countries, and have become one of the biggest challenges facing water resources in many parts of the world (Shiklomanov 1997:36).

Fortunately methods for monitoring change have improved as well. Many states and countries have moved beyond the conventional monitoring methods toward biological indicators of water quality. These are discussed in greater detail later. Groundwater resources are also important. Because groundwater is hidden from view, many pollution and contamination problems that affect supplies have been more difficult to detect and have only recently been discovered.

Condition Indicators of the Quality of Surface Waters

Information about water quality at the global level is poor and difficult to obtain for a number of reasons. Water quality problems are often local and natural water quality is highly variable depending on the location, season, or even time of day. Global criteria for water quality are, therefore, difficult to construct (Shiklomanov 1997:27).

However, there are many trends in the contamination of water supplies worldwide, and these have changed greatly over time. The main contamination problems 100 years ago were fecal and organic pollution from untreated human wastewater. The fecal contamination of water has been largely eliminated in most industrialized countries; however, organic matter pollution is still a problem in much of the world, especially in rapidly expanding cities in developing countries (Shiklomanov 1997:28). New pollution problems, particularly from agricultural runoff and industrial effluents, are increasing in both industrialized and developing countries. In rapidly industrializing countries, such as China, India, Mexico, and Brazil, untreated sewage and industrial wastes create substantial pressures on water quality that are much greater than the problems of the past (Shiklomanov 1997:27; UNEP/GEMS 1995:6).

A number of chemical, physical, and microbial factors negatively affect water quality (Taylor and Smith 1997; Shiklomanov 1997; UNEP/GEMS 1995). These include the following:

- Organic pollutants. Organic matter is a problem because it easily decomposes in water, consuming dissolved oxygen in the process. This often leads to the eutrophication and deoxygenation of waterways, with negative effects for aquatic life. The process also releases ammonium, which when converted to ammonia by natural chemical processes, is poisonous to fish. The primary sources of organic matter in lakes and rivers are wastewater from industrial plants and domestic sewage.
- Nutrients. Increased nutrient concentrations in freshwater, such as phosphorus and nitrates, can also cause eutrophication in lakes and rivers by decreasing the amount of oxygen available to aquatic life. This kills fish and other aquatic organisms. High levels of nitrates, when ingested in drinking water, restrict oxygen transport in the human bloodstream and can lead to illness.
- Heavy metals. These can be a severe problem because they accumulate in the tissues of fish and shellfish and are highly toxic. They also persist for long periods of time in freshwater ecosystems. Heavy metal pollution tends to be localized around industrial and mining centers.
- Microbial contamination. The contamination of water by bacteria, protists, and amoebae also pose threats to human health though the spread of infectious diseases. Fecal contamination from untreated sewage, for instance, leads to outbreaks of diseases that claim millions of lives each year.
- Toxic organic compounds. These include oil, petroleum products, pesticides, plastics, and industrial chemicals. All these are toxic to aquatic fauna and humans.
- Salinization. Increasing levels of salinity from overirrigation and groundwater overabstraction renders freshwater supplies undrinkable and kills crops.
- Acidification. Decreasing pH levels in rivers and lakes because of sulfuric deposition created by industrial activity kills fish and also leaches trace metals from soils, which has negative effects on human health.
- Suspended particles. These can be from either inorganic or organic matter. They degrade habitats of aquatic organisms and reduce water quality for drinking and recreational uses.
- Temperature: The thermal characteristics of water are crucial for aquatic life. Temperature determines the rate of chemical and biological processes, such as algal growth and decomposition of organic matter. Fragmentation of rivers by dams and reservoirs, as well as industrial uses such as hydropower and cooling plants, impact water temperatures.

CHEMICAL METHODS OF WATER QUALITY MONITORING

The amount of organic matter in freshwater systems around the world can be gauged by the global distribution of biochemical oxygen demand (BOD) measurements, seen in Figure 3. BOD increases with the amount of organic matter in water and gives an indication for the potential of algal growth and eutrophication *(see following note on the definitions of BOD and chemical oxygen demand, or COD)*. The BOD concentration of clean freshwater is normally around 2 mg/l, whereas values exceeding 5 mg/l usually indicate pollution (EEA 1994:45). The values in

Figure 3

Statistical Distribution of BOD by Continent, 1976–90

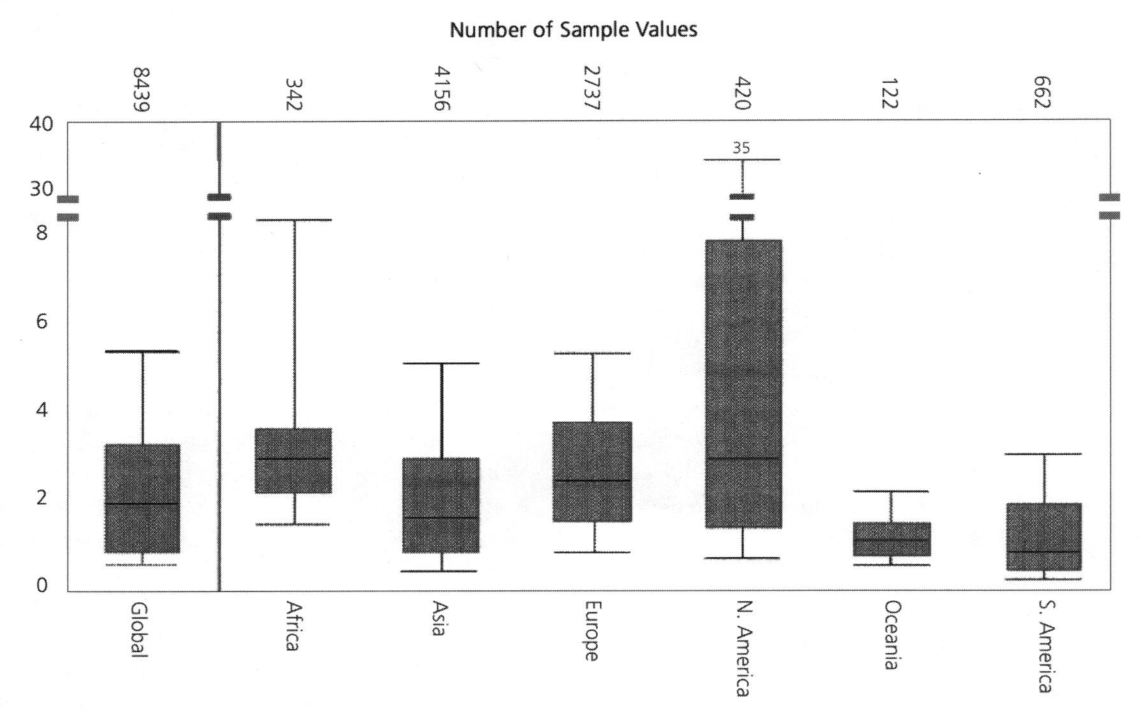

Source: UNEP/GEMS 1995.

Note: In Figure 3, the median value is represented by a horizontal line inside each of the gray boxes. The gray boxes represent the 25th and 75th percentiles, respectively. The two horizontal lines outside each gray box represent the 10th and 90th percentiles, respectively. Biochemical oxygen demand (BOD) and chemical oxygen demand (COD) are two methods widely used to measure the amount of organic pollution in wastewater and streams. BOD-5 is the amount of oxygen consumed by micro-organisms in a water sample over 5 days at 20 degrees Celsius. COD is the amount of oxygen consumed under specific conditions in the oxidation of organic and oxidizable inorganic matter contained in water. It is an indirect measure of the amount of oxygen used by inorganic and organic matter in water. In undisturbed rivers, BOD is typically less than 2 mg O_2/L and COD is less than 20 mg O_2/L. COD is a laboratory test based on a chemical oxidant and, therefore, does not necessarily correlate with biochemical oxygen demand (EEA 1994 and EEA 1998).

Figure 3 are based on the U.N.'s Global Environment Monitoring System (GEMS/WATER) that measured water quality from 1976 to 1990 at approximately 175 stations in 82 major river basins around the world (UNEP/GEMS 1995).

Although the BOD figures vary greatly, it can be seen that the highest organic matter concentrations were found in North American waterways. Values for Europe and Africa were also above the global median of 2 mg/l. Africa had a very wide range of values, which might represent greater amounts of pollution in the Nile; however more information is needed to give these figures greater clarity. But even in regions where organic matter concentrations are relatively low, such as Oceania and South America, eutrophication can be an issue. For example, the problems associated with eutrophication became apparent in Australia when the world's largest algal bloom spread along a 1,000-kilometer stretch of the Darling River in 1991 (SEAC 1996:7.49–50). This bloom caused the closure of water supplies for numerous communities along the river, forcing them to use costly alternative supplies (SEAC 1996:7.49–50). In this case BOD figures alone cannot adequately measure water quality because increased phosphorous concentrations are also a factor in noxious algal blooms.

These global data should be interpreted with caution, because the numbers of sampling points per basin vary in time and space. Continents are not represented equally by sampling points and the sample values cover a 15-year period with significant changes in human activities and management.

A more recent assessment of the water quality in 1,000 European rivers shows that, in the mid-1990s, 35 percent of rivers had BOD levels below 2mg/l while 11 percent were heavily polluted with levels of BOD greater than 5 mg/l. Of these heavily polluted rivers, 25 percent were in southern and eastern Europe (EEA 1999:172). Because there is no single standardized system of water monitoring and assessment in Europe for surface waters, comparisons between data from different countries are somewhat difficult (EEA 1998).

Improved treatment of wastewater from households and industry has brought about reductions in organic matter concentrations across Europe in the last 20 years. The largest reductions have been found in Western Europe, where the percentage of heavily polluted rivers has fallen from 24 percent in the late 1970s to 6 percent in the 1990s (EEA 1999:172). In southern Europe, BOD concentrations have started to improve slightly over the last 15 years, but in eastern Europe they have decreased from a high of 40 percent in the early 1980s to less than 28 percent in the mid-1990s (EEA 1999:173). Nordic countries have seen no major change in BOD concentrations since the late 1970s, and the number of river stations above natural levels is still below 5 percent (EEA 1999:172–173).

The level of nutrients in freshwater systems is an increasing problem worldwide (Shiklomanov 1997:34–36). Natural waters have very small concentrations of nitrates and phosphorous. But these nutrients increase with runoff from agricultural lands (especially intensively cultivated lands with large inputs of synthetic fertilizers) and urban and industrial wastewater, creating eutrophication and human health hazards. GEMS/WATER has collected the only global data on phosphorous and nitrate concentrations. They include data for major watersheds covering the period from 1976 to 1990, which carry the same data limitations mentioned for the GEMS/WATER BOD measurements. Of these globally monitored watersheds, the highest nutrient concentrations can be seen for sampling stations in Europe. Nitrate concentrations are higher in watersheds that have been intensively used and modified by human activity, such as the Weser, Seine, Rhine, Elbe, and Senegal. In South America, nitrate concentrations in the monitored watersheds are relatively low and follow human land use. The highest nitrate concentrations are found in the Uruguay watershed, where some of the most intensive agriculture on the continent is found. Nitrate concentrations are also greater in the Magdalena watershed of Colombia than in the less densely populated watersheds of the Amazon basin (UNEP/GEMS 1995:33–35). The nitrate concentrations in South America correspond to lower fertilizer application rates, compared to Europe. These low fertilizer application rates match an analysis of nutrient balances carried out for the PAGE agroecosystem study (Wood et al. 2000).

In Europe, for which more detailed and recent data are available, the concentrations of nitrates and phosphorous in rivers show distinct regional trends. Nitrogen loadings are the highest in areas with intensive livestock and crop production, especially in the northern parts of western Europe. Nitrogen concentrations are the lowest in Finland, Norway, and Sweden. Overall nitrate concentrations in the monitored European rivers have not changed significantly since 1980, despite lower nitrogen fertilizer application rates since the 1990s (EEA 1998:194–197; EEA 1999:176–177). Similarly, rivers in Finland, Norway, and Sweden have the lowest phosphorous concentrations, whereas areas from southern England across central and western Europe show the highest levels (EEA 1999:174). Even though phosphorous concentrations have decreased significantly since 1985, mostly because of wastewater treatment and the reduced use of phosphorous in detergents, it remains a problem in most regions of Europe (EEA 1999:174). Despite some positive trends, the overall state of many European rivers remains poor (EEA 1998:194–196).

Table 4 shows water quality data for the United States for the 1980s. Although data on stream water quality are continuously monitored, these are the latest aggregated figures published for all monitoring stations.

For the 1980–89 period, nitrate concentrations remained relatively stable, with nearly the same number of stations demonstrating upward trends as downward trends. This probably reflects the fact that nitrogen fertilizer use in the United States leveled off after steady increases in the 1970s. Fertilizer application rates increased for the period 1974–1981, and nitrate concentrations increased as well during that period. Average

Table 4

Trends in U.S. Stream Water Quality, 1980–89

Water Quality Indicator	NASQAN Stations Analyzed	Upward Trend in Concentration	Downward Trend in Concentration	No Concentration Trend
		Number of Stations		
Dissolved Solids	340	28	46	266
Nitrate	344	22	27	295
Total Phosphorous	410	19	92	299
Suspended Sediments	324	5	37	282
Dissolved Oxygen	424	38	26	360
Fecal Coliform	313	10	40	263

Source: Data are from the USGS National Stream Quality Accounting Network (NASQAN), quoted in CEQ 1995.

nitrate concentrations were greater in agricultural and urban areas than in forested areas (Smith et al. 1994: 122).

Trends in phosphorous concentrations in the United States showed greater improvement, with five times more states showing downward trends than upward trends. Decreases were more likely to be found in the East, Midwest, and the Great Lakes regions, while the majority of increases occurred in the Southeast (Smith et al. 1994: 124).

The decreased concentrations of phosphorous in streams and rivers in the United States is attributable to reduced phosphorous in laundry detergents and improved controls in wastewater treatment plants. The increased number of sewage treatment plants has also reduced the amount of nitrogen in the form of ammonium, which is toxic to fish. However, the sewage treatment process converts ammonium to nitrates that are still released into waterways. Thus, the total amount of nitrogen flowing into waterways has not necessarily decreased with a greater number of sewage treatment facilities (Mueller and Helsel 1996).

Water quality programs in most OECD countries have been effective in reducing many kinds of chemical pollution in their waterways. The biggest reductions have been from point source pollutants. For example, in the United States from 1972 to 1992 the amount of sewage treated at wastewater treatment plants increased by 30 percent, yet BOD measurements of waters near these plants declined by 36 percent (CEQ 1995: 229). However, national programs have not been effective in reducing nonpoint source nutrients, sediments, and toxics that come from agriculture, urban and suburban stormwater runoff, mining, and oil and gas operations (NRC 1992:47;EEA 1999:178).

Similarly the existing data tell us little about the biological characteristics of inland waters because historically, water quality monitoring has focused on the measurement of chemical parameters. Although such information has provided important and predictive tools for evaluating water quality, the monitoring of biological indicators to assess the health of freshwater ecosystems is now also recognized as an important component of water quality monitoring programs. Beginning in the 1980s and increasingly in the last 10 years, a greater number of states in the United States and other countries are including biological monitoring as an important part of their overall water quality monitoring programs.

BIOLOGICAL METHODS OF WATER QUALITY MONITORING

Measurements of water quality based exclusively on chemical or physical properties are often used as surrogate measures of biological quality. But a reliance on these measures alone risks missing many important biological characteristics, such as habitat alteration and species composition. Furthermore, improvements in chemical parameters in many cases will not lead to increases in biological integrity by themselves. Biological monitoring goes beyond the conventional measures of water quality to address questions of ecosystem function and integrity.

Early attempts to apply biological criteria to water quality monitoring were largely qualitative, narrative descriptions based on a single-dimension metric, such as general species richness (Yoder and Rankin 1998). These single-variable measures did not give an adequate understanding of the complex biological interactions in aquatic ecosystems. To address these problems, the biologist James Karr developed the Index of Biotic Integrity (IBI) for freshwater habitats in 1981 (Karr and Chu 1999:2). The goal of the IBI is the integration of data about fish species, populations, and assemblages into a single comparative numeric indicator. Karr's IBI used 12 quantitative metrics based on 3 categories: species richness and composition, trophic composition, and fish abundance and condition. It was first applied to streams and rivers of the midwestern United States.

A number of states in the United States have modified and adjusted the IBI, with Ohio having one of the most comprehensive programs for the assessment of freshwater biological integrity. After seven years of developing its IBI, Ohio formally adopted the index in 1990 and now uses biological monitoring as an established part of its water quality assessment programs (Yoder and Rankin 1995). Ohio's IBI modified some of the original 12 metrics and adjusted them to the state's natural landscape variability (Yoder and Rankin 1995). Whereas the standard IBI concentrates on fish diversity, abundance, and community structure, Ohio has a separate index for invertebrate species, called the Invertebrate Community Index (ICI), which detects additional trends and ecosystem conditions (DeShon 1995).

Less comprehensive but still effective indices have been developed in other states to measure biotic integrity. The Florida Department of Environmental Protection created the Stream Condition Index (SCI) to measure the effects of nonpoint source pollution on biological integrity in the state, using macro-invertebrates as indicator species. The SCI metrics include the total number of taxa, the total number of insect taxa (mayflies, stoneflies, and caddisflies), the number of midge larvae taxa, the percent of dominant taxon, the percent of diptera, the percent of filterers, and the weighted sum of tolerant and intolerant species. With increasing disturbance, species tolerance and richness are expected to decrease, while species composition will change differentially depending on the species. As with other IBIs, the Florida SCI scores reflect these potential changes. State officials continue to modify the SCI (Barbour et al. 1996).

The IBI used in Ohio requires data collection and analysis from hundreds of sites each year, which means that time and money must be spent to obtain IBI scores and recalibrate reference site values. Indices that use macro-invertebrates alone are less expensive because the costs of data collection are not as high for invertebrates as for fish. However, using more than one taxa increases the amount of information available to resource managers. Ohio officials have found, however, that biological surveying is cost-competitive with water chemistry surveys and bioassays (Yoder 1991:101–102). Chemical assessments, for example, have lower unit costs per sample but many more of them are needed. For a single stream in Ohio, chemical assessment in the late 1980s cost up to US$32,400 for 90 samples, compared with US$22,000 for 12 biological surveys. Nine bioassays cost between US$16,000 and US$28,000 (Yoder 1991:101).

Both chemical and biological criteria serve as important components in assessing water quality. Ohio compared biological and chemical assessments in 625 water body segments throughout the state in 1990. In 49.8 percent of the samples, biological impairment was detected with either an IBI or ICI in which no chemical impairment was measured. Both biological and chemical impairment were found in 47.4 percent of the samples, whereas only 2.8 percent of the samples found chemical impairment without any corresponding measured biological degradation. This comparison demonstrated the importance of using biological criteria in addition to chemical indicators of water quality. In a majority of cases in which only biological impairment was measured, the reasons for decreased biological integrity (increased organic matter, habitat modification, siltation) could not be measured using chemical criteria alone (Yoder 1991:98).

A similar approach to evaluating ecosystem conditions that uses whole-watershed metrics rather than site-specific variables within freshwater systems alone is the Watershed Index of Biotic Integrity (W-IBI). This index was developed to evaluate 100 watersheds in the Sierra Nevada Mountains of California (Moyle and Randall 1998:1320). The W-IBI uses such variables as the number of dams, reservoirs, and water diversions, along with the percentage of the area with roads in each watershed to assess condition. Data on native and introduced fish species— as well as on other taxa, such as native frogs, which are highly sensitive to disturbance and have completely disappeared from many watersheds— were also used as indicators. Indices were scored on a scale of 20 to 100, with 80 to 100 being excellent. Only 7 out of 100 watersheds were ranked in the excellent category. Watersheds that scored poorly were those at lower elevations where stream channels had been highly modified by dams, diversions, agriculture, or urbanization, and high-elevation streams where native frogs had declined and introduced predatory fish were present in formerly fishless areas. The W-IBI gives a good watershed-level estimation of biotic integrity and health, but sacrifices some local-level information in the process. For example, within a poorly ranked watershed, there can be streams that have very high biotic integrity, but this will not show up in the W-IBI classification (Moyle and Randall 1998:1323–1325).

The IBI has also been applied outside the United States. For example, an IBI has been developed for the Seine River basin in France (Oberdorff and Hughes 1992). Analysis of these data showed that IBI scores decreased through time, reflecting increasing amounts of pollution and habitat disturbance since the 1960s. IBI values varied longitudinally as well, decreasing in all years in a downstream direction because of an increasing amount of disturbance, a trend that became more marked after 1967.

Another application of the IBI was also used to assess the biological integrity of rivers in India (Ganasan and Hughes 1998) and Mexico (Lyons et al. 1995). The study in India, for example, examined two rivers that had large quantities of untreated waste and toxic heavy metals (Ganasan and Hughes 1998:367). The IBI scores, based on 1986, 1989, and 1991 data, increased downstream from cities and towns, reflecting a gradual recovery of biotic integrity with increasing distance from pollution sources. Nonnative fish species comprised between 4 and 55 percent of individuals at sites where water flow was restricted by an impoundment, compared with 1 to 2 percent of individuals at the least disturbed sites. One of the main conclusions of this study was that the original IBI format could be adapted to Indian rivers, despite an overlap of only two families and no species between the midwestern United States and the Indian study site. However, further evaluations of this trial study need to be undertaken before the Indian IBI can be widely used in an effective and cost-efficient manner (Ganasan and Hughes 1998:378–379).

An alternative to the IBI for making biological assessments of water quality is multivariate statistical analysis of aquatic communities to make predictions of species composition in different sites. Multivariate models have been developed that predict the number of macro-invertebrate freshwater fauna expected to occur at a given site in the absence of environmental stress. The observed invertebrate fauna is then compared to the expected fauna based on statistical methods, and the ratio of observed to expected fauna is used to classify the health of a site. Unstressed sites should have observed/expected ratios that are close to one, whereas stressed sites will have lower ratios.

Examples of freshwater biological monitoring programs that use multivariate statistical methods can be found in the United Kingdom (Wright 1995), Australia (Marchant et al. 1997), and the state of Maine in the United States (Davies et al. 1995). In most cases the sampling strategies used to assess biological communities are the same as those used in IBIs, and the two methods generally produce similar results. Multivariate methods of analysis have the potential to produce accurate predictions of species compositions at unsampled sites based on the correlation between reference site data and measured environmental variables of water quality (Marchant et al. 1997:664).

But more data must be collected than with the IBI to achieve better results.

Condition Indicators of Groundwater Quality

The major sources of groundwater pollution are leaching of pollutants from agriculture, industry, and untreated sewage, saltwater intrusion caused by overabstraction of groundwater (which was discussed in a previous section), and natural hydrogeochemical pollution. Because of the relative inaccessibility and slow movement of the resource, polluted groundwater is very difficult to purify (UNEP 1996:22). Once pollutants enter a groundwater aquifer, the environmental damage can be severe and long lasting, partly because of the very long time needed to flush pollutants out of the aquifer (UNEP 1996:14). Because it is primarily used for drinking water, groundwater pollution from untreated sewage, intensive agricultural, solid waste disposal, and industry can cause serious human health problems (Shiklomanov 1997:42). Global data on the quality of groundwater resources is lacking. Even where available, data usually are not comparable because of the different measures and standards used, which vary by country (Shiklomanov 1997:42; Scheidleder et al. 1999:11; Foster, personal communication, 2000). However, there is evidence that groundwater contamination from fertilizers, pesticides, industrial effluents, sewage, and hydrocarbons is occurring in many parts of the world.

Because the source of groundwater pollution is determined by local conditions, and these vary widely, we have selected particular cases to illustrate pollution problems affecting groundwater resources around the world. The selection of cases is based mostly on data availability. This overview of groundwater quality is by no means comprehensive and portrays trends for only some pollutants and certain regions of the world.

NITRATE POLLUTION

Nitrate pollution of groundwater sources is a problem in both industrialized and industrializing countries (UNEP 1996:28). Nitrate pollution can come from agricultural, urban, or industrial sources; untreated sewage is a major source of nitrate pollution in many parts of the world (UNEP 1996:22). Natural nitrate levels in groundwater are low, usually around 10 mg/l of nitrate (Scheidleder et al. 1999:19). Nitrate moves slowly in soil and groundwater, so there is typically a time lag of 1 to 20 years from the time of pollution until its detection.

In China, nitrogen fertilizer consumption increased sharply in the 1980s and now equals application rates in western Europe. Fourteen cities and counties in northern China, covering an area of 140,000 square kilometers, were sampled to assess the extent of nitrate contamination of groundwater supplies (Zhang et al. 1996: 224). This area had over 20 billion cubic meters of groundwater withdrawals in 1980, mostly for irrigation but with large amounts used for drinking water supplies as well (UN 1997:29). Over one half of the sampled areas had nitrate concentrations that were above the allowable limit for nitrate in drinking water. The majority of these were smaller towns and cities (10,000 to 100,000 in population) that were surrounded by agricultural areas with high fertilizer application rates and that depended on groundwater for the majority of their drinking water supplies (Zhang et al. 1996:227). Pollution of groundwater supplies from synthetic fertilizer application is also a problem in parts of India. Groundwater samples in the states of Uttar Pradesh, Haryana, and Punjab were found to have between 5 and 16 times the prescribed safe amount of nitrate, with one site in Haryana almost 30 times the prescribed limit. Groundwater in these areas is also being progressively depleted because of overabstraction for irrigation (TERI 1998: 214–215).

Nitrate pollution of groundwater supplies is also a problem in industrialized countries, as can be seen from a recent European assessment of groundwater resources (Scheidleder et al. 1999). Twenty-two countries reported regional-level data on nitrate pollution. The number of sampling sites varied widely from one region to another. In 50 of the reported regions, nitrate levels exceeded 25 mg/l in at least a quarter of the total samples. The levels exceeded 25 mg/l in half of the samples in an additional 13 regions. In some regions of France, the Netherlands, and Slovenia, nitrate concentrations exceeded 50 mg/l in 67 percent of sampling sites. In Poland and Moldova, groundwater wells with nitrate concentrations in excess of 45 mg/l could be found across all parts of both countries (Scheidleder et al. 1999:54).

In general, the risk of nitrate pollution for groundwater supplies is directly related to the amount of fertilizers or other nitrogen inputs to the land, and the permeability of the soils through which nitrogen is leached. In the United States, groundwater provides drinking water for more than onehalf of the nation's population (UNEP 1996:12). Data collected by the USGS NAWQA program demonstrate that nitrate concentrations in groundwater exceeding the recommended level of 10 mg/l are significantly greater in aquifers that have high nitrogen inputs and are most vulnerable to leaching (Nolan et al. 1998). Nevertheless, little information exists about groundwater supplies at the national level in the United States, and data are especially poor for ammonium and phosphorous concentrations.

The USGS has begun a more comprehensive program of groundwater quality monitoring, but several more years of sampling will be needed before enough data will be collected to analyze national trends (Mueller and Helsel 1996). A preliminary analysis of nitrate in U.S. groundwater, however, shows that high concentrations in shallow groundwater are widespread and closely correlated with agricultural areas, yet no regional patterns could be discerned (USGS 1999:41). The application rates of synthetic nitrogen fertilizer can be used as a general indicator of groundwater quality because increasing levels of agricultural intensification usually mean overapplication of fertilizers and subsequent leaching into groundwater supplies (Scheidleder et al. 1999:21).

NATURAL (HYDROGEOCHEMICAL) POLLUTION

Natural groundwater contamination in the Indian state of West Bengal and in Bangladesh currently threatens the lives of millions of people. In West Bengal, 7 districts covering an area of over 37,000 square kilometers are contaminated with arsenic, with an estimated 1.1 million people drinking from contaminated wells. A survey found that 200,000 people were suffering from arsenic-related diseases, but the numbers could potentially be much higher because the total population of the affected area is over 9.5 million people (Mandal et al. 1996:976). Naturally occurring arsenic also affects wells across Bangladesh; as many as one million wells in Bangladesh and west Bengal might be contaminated. Ironically, many of these wells were drilled to provide a safe drinking water alternative to heavily polluted surface water supplies. (Nickson et al. 1998: 338).

Capacity of Freshwater Systems to Provide Clean Water

Surface water quality has improved in most OECD countries during the past 20 years, but nitrate and pesticide contamination remain persistent problems. Data on water quality in other regions of the world are sparse, but water quality appears to be degraded in almost all regions with intensive agriculture and rapid urbanization. Unfortunately, little information is available to evaluate the extent to which chemical contamination has impaired freshwater biological functions. However, incidents of algal blooms and eutrophication are widespread in freshwater systems all over the world—an indicator that these systems are profoundly affected by water pollution. In addition, the massive loss of wetlands at a global level has greatly impaired the capacity of freshwater systems to filter and purify water.

Nitrate pollution of groundwater supplies in northern China and India is likely to remain a serious problem for years to come. Increasing populations mean that agricultural productivity has to increase to meet growing needs, yet in China the amount of arable land is slowly decreasing due to urbanization. It is estimated that application rates for nitrogen fertilizer will double or triple in the next 30 years (Zhang et al. 1996: 223). Not only are there negative health effects associated with nitrate pollution of groundwater, but overapplication has serious economic costs as well. For one 115,000-hectare region in the northern province of Shaanxi, economic losses because of excessive application of nitrogen fertilizer were estimated to be US$13 million (Emteryd et al. 1998:443).

Water Quality Information Status and Needs

Because the quality of water is one of the most critical factors affecting the quality of life on Earth, monitoring water quality is an urgent and important task. Unfortunately, there have been few sustained programs for the global monitoring of the quality of water, with the result that information is highly localized and far from complete.

One of the only global attempts at water quality monitoring has been the UNEP GEMS/WATER program that examined data from 82 major river basins worldwide over a period of a decade and a half. This program gathered data on a variety of water quality issues, including nutrients, oxygen balance, suspended sediments, salinization, microbial pollution, and acidification. Yet, the number of monitored watersheds were too sparse and the frequency and type of measurements were too inconsistent to paint a comprehensive picture of global water quality trends. Further studies of this kind need more comprehensive and systematic data collection, and monitoring should be carried out indefinitely so that long-term trends can be analyzed. Data needs are especially critical for developing countries, which often do not have strong national monitoring programs, yet, face serious water quality problems.

Surface water monitoring programs are relatively well developed in most OECD countries, where many different physical and chemical measures are monitored. Data for groundwater are less reliable in most cases, and much more information needs to be gathered in all countries on groundwater quality. Even in the United States, efforts by the USGS to monitor groundwater quality have begun only in recent years, and current monitoring covers only a portion of the country's land area.

In developing countries, the situation is much worse. There are a few localized studies of groundwater quality for many developing countries, but no country has a comprehensive program for groundwater monitoring. A global program for addressing groundwater quality issues could be built around a few indicator variables that are most likely to adversely affect human health (such as nitrates, salts, and toxic chemicals), with frequent monitoring carried out in areas of heavy dependence on groundwater use.

One of the most important issues that must be addressed in any global program of water quality monitoring is the need for more biological monitoring. It has been shown that chemical monitoring alone fails to identify many instances in which freshwater ecosystems are stressed or threatened by nonchemical factors. Declines of biological integrity can have adverse effects on human welfare as well, through declining fish and shellfish stocks. Biological monitoring is needed to give greater information about overall ecosystem health and integrity.

However, unlike programs for the monitoring of chemical and physical parameters, biological monitoring cannot easily be carried out on a global scale. This is because the information needed to conduct useful indicators of biological integrity cannot be found in single-measure indices. Information about freshwater biotic integrity is highly dependent on local circumstances, such as the numbers and types of organisms and the ways in which freshwater biological communities are structured. No single IBI-type measure could be constructed for the entire world because IBIs require locally calibrated reference sites to be useful. Unlike conventional chemical measures of water quality, biological indices are difficult to construct even at the national level, depending on the size of the country. In the United States, where several nationwide programs for measuring chemical aspects of water quality are underway, biological monitoring is conducted at the state level, and it probably will never be coordinated at a higher level of administration.

One of the biggest challenges in future global water monitoring programs is the integration of chemical and biological measures of water quality. Although the former can be carried out at national and even continental scales, the latter must be approached at the local and regional level. Yet, both have important messages to tell us about water quality and ecosystem health.

FOOD—INLAND FISHERIES

Status and Trends in Inland Fisheries

Fish are a major source of protein and micronutrients for a large part of the world's population, particularly the poor (Bräutigam 1999:5). Inland fisheries in rivers, lakes, and wetlands are an important source of this protein because almost the entire catch gets consumed directly by people—there is practically no bycatch or "trash" fish in inland fisheries (FAO 1999a:1). The population of Cambodia, for example, obtains roughly 60 percent of its total animal protein from the fishery resources of the Tonle Sap alone (MRC 1997:19). In some landlocked countries, this percentage is even higher. Inland fisheries in Malawi provide about 70–75 percent of the total animal protein for both urban and rural low-income families (FAO 1996: 3).

The catch from inland fisheries totaled 7.7 million metric tons in 1997, or nearly 12 percent of all fish directly consumed by humans from all inland and marine capture fisheries (FAO 1999b:7). Inland fisheries landings are comprised mostly of freshwater fish, although molluscs, crustaceans, and some aquatic reptiles also are caught and are of regional and local importance (FAO 1999b:9). The catch from inland fisheries is believed to be greatly underreported—by a factor of two or three (FAO 1999a:4). Asia and Africa are the two leading regions in inland capture fish production (*see Figure 4*).

In fact, 5 of the top 10 producing countries are in Asia. China is the most important inland capture fisheries producer, with 24 percent of the world total (*see Figure 5*) (FAO 1999a:7).

According to the FAO, most inland capture fisheries that depend on natural production are being exploited at or above their maximum sustainable yields (FAO 1999a:25). Globally, inland fisheries landings increased at 2 percent per year from 1984 to 1997, although in Asia the rate has been much higher—7 percent per year since 1992. This increase has occurred partially as a result of efforts to raise production above natural levels through fisheries enhancements and to eutrophication of inland waters from agriculture runoff and certain types of industrial effluents (FAO 1999a:6–7; Kapetsky, personal communication, 1999). Asia, in particular China, which is the leading country in inland fish production, produces 64 percent of all the inland fish catch –much more than would be expected given the available freshwater area if compared with other regions of the world. This, plus the large number of reservoirs in China, points to the strong use of fishery enhancements, such as stocking, to increase yields (FAO 1999a:7). These enhancements, however, can seriously affect the condition and long-term functioning of freshwater ecosystems (*see Biodiversity section*).

Freshwater aquaculture is now more important than inland capture fisheries, with total production estimated to be 17.7 million metric tons in 1997 (FAO 1999a:6). In freshwater sys-

Figure 4

Inland Capture Fisheries by Continent, 1984–97

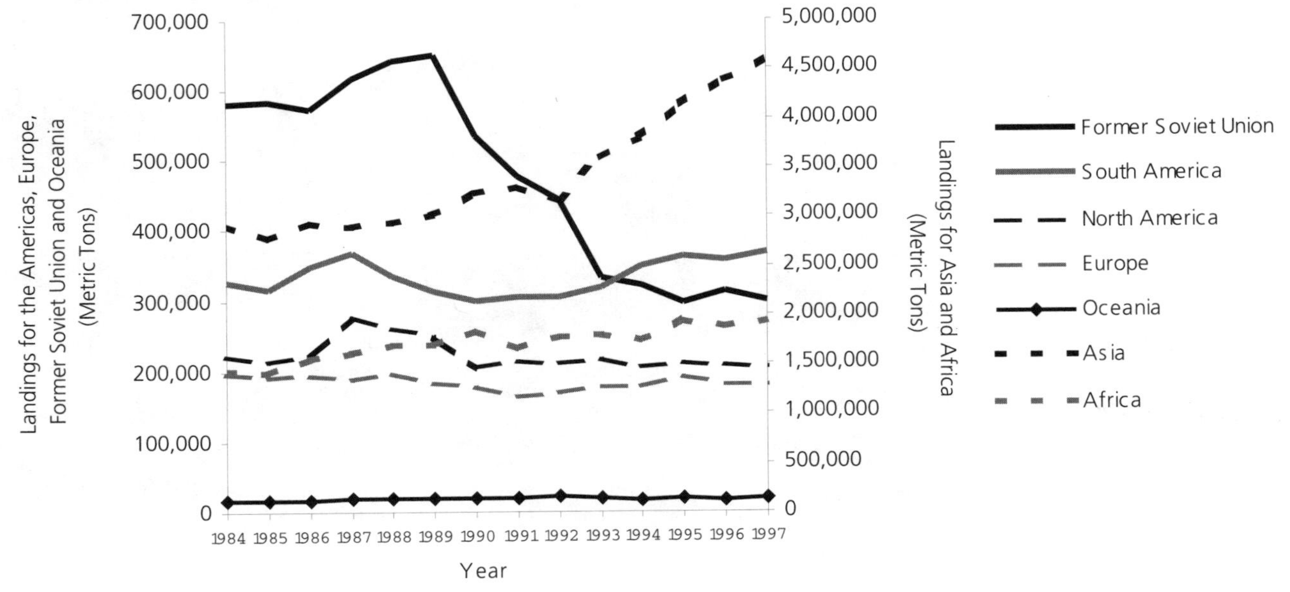

Source: FAO 1998.

tems, finfish aquaculture represents 99 percent of production (FAO 1999b:12) and it has been increasing rapidly for the last 15 years, with an average annual rate of increase of 11.8 percent since 1984 (FAO 1999b:81). *(See Figure 6)*.Both, marine and freshwater aquaculture have become a critical source of food for the world's population, providing 30 percent of the fish for human consumption in 1997 (FAO 1999b:7). Excluding the production of aquatic plants, more than 60 percent of aquaculture production is freshwater fish or fish that migrate between fresh and saltwater (FAO 1998). Asia, and China in particular, dominate aquaculture production in terms of volume. A large part of their production is carp (FAO 1999a:10).

In more developed regions such as Europe and North America, freshwater fish consumption has decreased over time and recreational fishing is replacing inland food fisheries (FAO 1999a:8). Recreational fishing contributes a significant amount of revenue to the economies of some countries. For instance, anglers in Canada spend 2.9 billion Canadian dollars a year in goods and services directly related to fishing (McAllister et al. 1997:12) and in 1996, anglers in the United States spent US$447 million just on fishing licenses alone (FAO 1999a:42). Recreational fishing also contributes to the food supply in the same way that subsistence fishing does because anglers usually consume what they catch, although there has been a trend recently to release fish after they are caught (Kapetsky, personal communication, 1999). Currently recreational catch is estimated at about 2 million metric tons per year (FAO 1999a:42). This positive trend in recreational fishing is not limited to developed countries, but has also been taking place in many developing countries.

Recreational and food fisheries in many countries are maintained by fishery enhancements, particularly species introductions and stocking (FAO 1999a:26). Enhancements to increase recreational fisheries are more prevalent in North America, Europe, and Oceania, while enhancements to increase food production are more prevalent in Asia, Africa, and South America (FAO 1999a:28–32). Because of the enhancements, inland fisheries have become an important factor in food security and income generation in many regions of the world (FAO 1999a:26).

It should be mentioned that some types of aquaculture, particularly the farming of such carnivorous species as salmon, depend in part on marine capture fisheries because fish meal, usually from small pelagic fish, contribute to many compound feeds used in aquaculture (Naylor et al. 2000:1017). Aquaculture operations, depending on their design and management, can also contribute to habitat degradation, pollution, introduction of exotic species, and the spread of diseases through the introduction of pathogens (Naylor et al. 2000:1017).

Pressures on Inland Fishery Resources

The principal factor threatening inland capture fisheries is the loss of fish habitat and environmental degradation (FAO 1999a:19). Most of the world's freshwater systems have been modified to a certain degree. Rivers have been physically altered by dams and reservoirs, or channeled and constrained to prevent and control floods, with the consequent loss of riparian

Figure 5

Top Ten Producing Countries, 1997

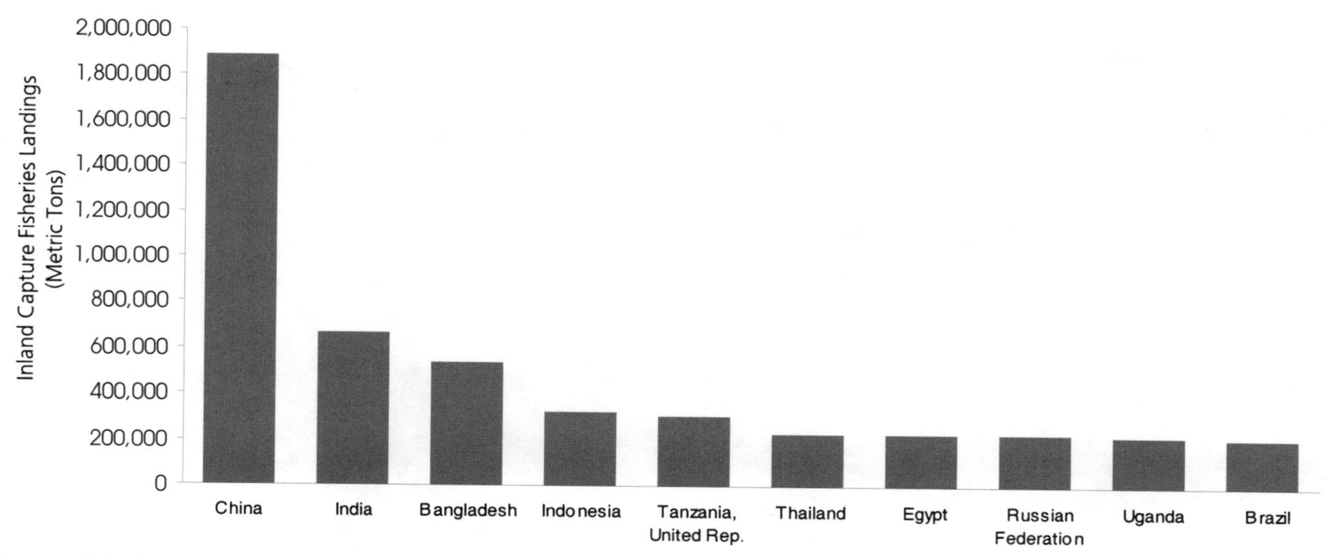

Source: FAO 1998.

habitat. Wetlands, some of the most productive ecosystems in terms of fish production, have been and continue to be drained throughout most of the world. In such areas as the Mekong River basin and other parts of Asia, overfishing and destructive fishing practices also contribute to the decline in inland fisheries production (FAO 1999a:19). In addition, nonnative species introduced into lakes, rivers, and reservoirs, either accidentally or for food production or recreational fishing, affect the composition of the native aquatic communities. The effect is to sometimes increase levels of production and sometimes decrease them. Introduced species often are predators or competitors, and they may spread new diseases to the native fauna, sometimes with severe consequences, as seen in the case of Lake Victoria (*discussed in the prologue*).

When different stresses affect an aquatic system over time, the species composition changes considerably. This change in species assemblages has been observed in many rivers in developed regions of the world. These changes in wild stocks serve as an indicator of the changes of the environmental quality of the system. In Europe and North America, dams, channels, and interbasin transfers have dramatically altered most rivers. Rivers and lakes have been heavily polluted from sewage, industrial sources, and agricultural runoff. There have also been numerous species introductions in the rivers, lakes, and reservoirs of these two regions. Fish assemblages have, therefore, changed dramatically and many species have become threatened, rare, or extinct (Arthington and Welcomme 1995:57, 58). The recent awareness of environmental issues, and a growing appreciation of the importance of freshwater systems for recreational and aesthetic purposes, are making restoration and integrated multiple-use projects more prevalent in Europe and North America (FAO 1999a:20).

In South America and Africa, dams and reservoirs have also significantly modified rivers, especially in the larger basins. Pollution, particularly sewage contamination, is increasing close to urban centers and heavy chemical pollution is found in mining and other industrial areas. Siltation from deforestation and resulting eutrophication is also increasing in many lakes, particularly in Africa. Localized overexploitation from uncontrolled fisheries and a change in species composition from native to introduced species are becoming increasing problems in many lakes in Africa (Arthington and Welcomme 1995:59, 60; FAO 1995b:7–16, 18–22).

Asia depends heavily on inland fisheries for food production. In China, most rivers have been extremely altered and polluted, whereas in Southeast Asia rivers are less modified and still retain their natural flooding patterns. These flooding regimes and the associated flood plains are at risk of being lost to development activities, such as expanded agriculture or dam construction, which are being driven by an increase in population and the consequent growing demand for food, water, and electric power. Losing these resources, which are self-sustaining fish production systems, could have a drastic impact on food security for the rural population in the region (FAO 1995b:26–27).

As environmental degradation increases, and the demand for inland fish continues to grow, the only way to maintain inland fisheries production is through fisheries enhancements or

Figure 6

Global Inland Capture and Freshwater Aquaculture Growth, 1984–97

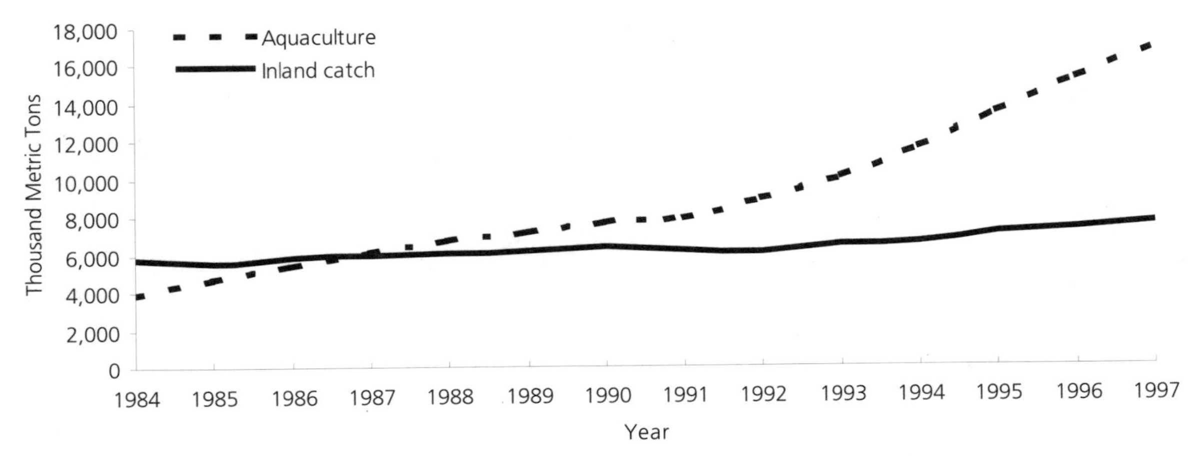

Source: FAO 1998.

the restoration of rivers and aquatic habitats. Restoration and rehabilitation of rivers is usually a costly process and is only practiced where there is public support and available finances (FAO 1999a:43). In contrast, fisheries enhancements are practiced in almost every country with inland fisheries resources (FAO 1999a:26). The enhancements include various techniques, such as the introduction of new species, stocking, and fertilization. In most cases, they are meant to increase food or recreational fish production or to control pests, but often they put added stress on freshwater species communities. Many of the enhancements pose some threats to the condition of the ecosystem, mostly through loss of species and genetic diversity, and changes in species assemblages (*see Biodiversity section for additional discussion of these issues*).

Condition Indicators of Fish Production

Assessing the actual condition of inland fisheries, rather than overall threats to the fisheries at the global or regional level, is difficult partly because of the paucity of reliable and comprehensive data on fish landings. The FAO has collected data on inland capture fisheries since 1984. However, the data collection and reporting have several important weaknesses that make it difficult to assess the state of the resource (*see the following section on Information Status and Needs for more detailed discussion on the limitations of inland fisheries statistics*).

Fish communities also change considerably over time because of the combination of stresses. In general, during the first phase of exploitation of a fishery, fishers start by targeting those large and most valuable species. As these species are overfished, they are replaced by smaller fish, which in turn become the target of the fishery (Arthington and Welcomme 1995:53). Fishing pressures in combination with other stresses make the focus of the fishery change regularly and, therefore, catch statistics of the present quality are of limited value to assess the condition of the system.

Nevertheless, harvest and trend information exists for certain well-studied fisheries. This study looked at two indicators of the condition of inland fishery resources. One indicator comprises historical trends in catch statistics and changes in fish fauna composition for selected well-studied rivers, lakes, and inland seas. The second is the recent trends in catch statistics from the FAO database on inland fisheries (1984–97).

HISTORICAL CHANGE IN FISH CATCH AND SPECIES COMPOSITION

An indicator of the condition of freshwater systems is the number and diversity of fish species that they support, especially compared with historical data. Because fisheries and species assemblages are influenced by many stressors, and because the impacts from these stressors are not immediately felt, it is necessary to examine the change in fish catch and species composition over a long period of time in order to assess the condition of a freshwater system. Unfortunately, historical fish harvest data have not been systematically collected for many rivers and lakes around the world. Therefore, this study relied on available commercial fisheries data and species composition for specific water bodies. Table 5 shows the changes in freshwater fish species and commercial fisheries since the early or mid-1900s for selected rivers, lakes, and inland seas. Without exception, each of these major fisheries has experienced dramatic declines during the 20th century.

For some rivers in which fisheries have been tracked more systematically, catch statistics help assess the capacity of the

Table 5

Changes in Fish Species Composition and Fisheries for Selected Rivers, Lakes, and Inland Seas

River	Change in Fish Species and Fishery	Major Causes of Decline	Main Goods and Services Lost
Colorado River (United States)	Historically native fish included 36 species, 20 genera, and 9 families. Of these, 64% were endemics. Current status of species under the Endangered Species Act: 2 extinct, 15 threatened or endangered, 18 proposed for listing or under review.	Dams, river diversions, canals, and loss of riparian habitat.	Loss of fisheries and biodiversity.
Danube River	Danube River fisheries have changed dramatically since the early 1900s. Danube sturgeon fishery has almost disappeared, and current fisheries are maintained through aquaculture and introduction of nonnative species.	Dams, creation of channels, pollution, loss of floodplain areas, water pumping, sand and gravel extraction, and nonnative species introductions.	Loss of fisheries, loss of biodiversity, and change in species composition.
Aral Sea	Of 24 fish species, 20 have disappeared. The commercial fishery that used to have a catch of 40,000 tons and support 60,000 jobs is now gone.	Water diversion for irrigation, pollution from fertilizers and pesticides.	Loss of important fishery, loss of biodiversity. Associated health effects caused by toxic salts from the exposed lakebed.
Rhine River	Forty-four species have become rare or disappeared between 1890 and 1975. Salmon and sturgeon fisheries are gone, and the yield from the eel fisheries has declined, even though it is maintained by stocking.	Dams, creation of channels, heavy pollution, and nonnative species introductions.	Loss of important fishery, loss of biodiversity.
Missouri River	Commercial fisheries declined by 83% since 1947.	Dams, creation of channels, and pollution from agriculture runoff.	Loss of fishery and biodiversity.
Great Lakes	Change in species composition, loss of native salmonid fishery. Four of the native fish have become extinct, seven others are threatened.	Pollution from agriculture and industry, nonnative species introductions.	Loss of fishery, biodiversity, and recreation. Contamination of fish leading to fish advisories and human health problems.
Illinois River	Commercial fisheries declined by 98 percent in the 1950s.	Siltation from soil erosion, pollution, and eutrophication.	Loss of fishery and biodiversity.
Lake Victoria	Mass extinction of native cichlid fish. Changes in species composition and disappearance of the small-scale subsistence fishery on which many local communities depended.	Eutrophication, siltation from deforestation, overfishing, and nonnative species introductions.	Loss of biodiversity and local artisanal fishery.
Pearl River (Xi Jiang)	In the 1980s the yields in commercial fisheries had dropped to only 37 percent of the yield levels of the 1950s.	Overfishing, destructive fishing practices, pollution, and dams.	Loss of fishery.

Sources: Carlson and Muth 1989; Bacalbaça-Dobrovici 1989; Postel 1995; Lelek 1989; Hughes and Noss 1992; Sparks 1992; Kauffman 1992; Missouri River Coalition 1995; and Liao et al. 1989.

Figure 7

Commercial Landings of Salmon and Steelhead from the Columbia River, 1866–1998

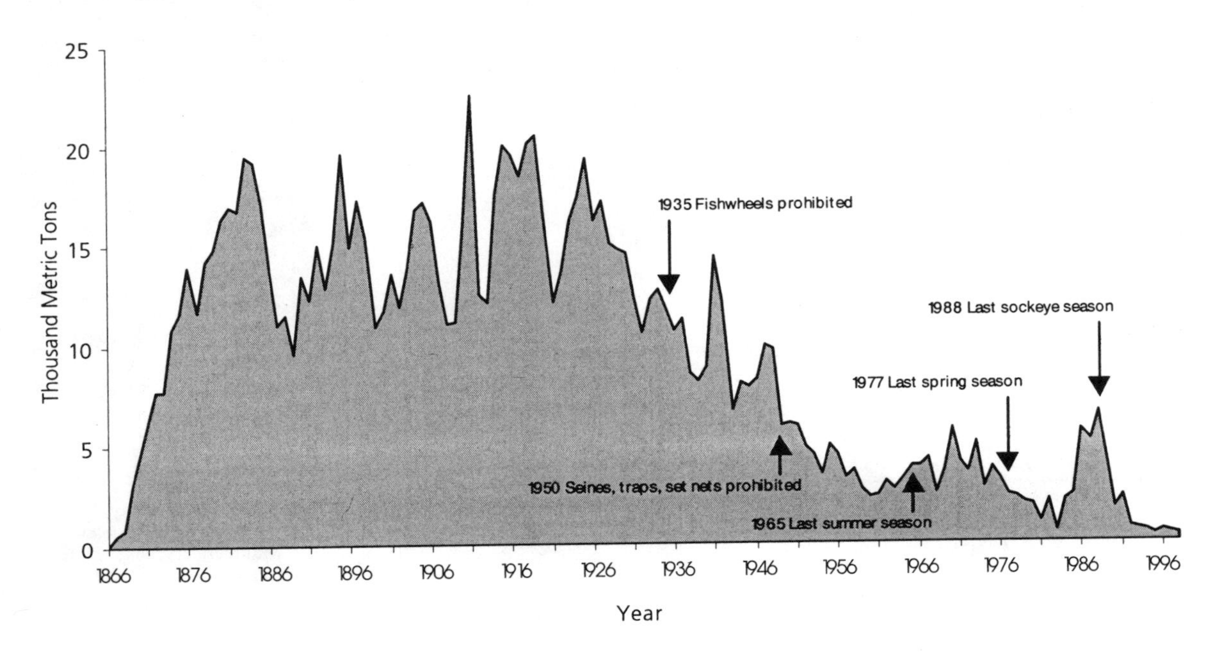

Source: Washington Department of Fish and Wildlife and Oregon Department of Fish and Wildlife 1999.

systems to provide a particular good—for example, salmon and steelhead production from the Columbia River (*see Figure 7*). Commercial landings of salmon and steelhead have been falling since the 1930s, with commercial fishing being prohibited for particular seasons after 1965, 1977, and 1988 (Washington Department of Fish and Wildlife and Oregon Department of Fish and Wildlife 1999).

Because fish are good indicators of the condition of the aquatic system, the condition of the fishery serves also as an indirect measure of the condition of the system as a whole. As can be seen in Table 5 and Figure 7, there has been considerable decline in landings of commercial fisheries as well as changes in species composition and biodiversity loss for these selected cases. In all instances, the cause for these declines is a combination of multiple stresses. Even though this is a small sample of cases, it illustrates what is occurring in other aquatic systems around the world. As ichthyologists Harrison and Stiassny put it, "although the precise degree of freshwater impoverishment remains to be fully documented, there can be little doubt that the losses are already great" (Harrison and Stiassny 1999:271).

RECENT TRENDS IN CATCH STATISTICS

A somewhat different picture of the condition of inland fisheries is provided by data from the FAO, which has been used as a diagnostic for the changes in inland fisheries production over the last 14 years. By analyzing catch statistics, the FAO found positive trends in inland capture fisheries in South and Southeast Asia, Central America, and parts of Africa and South America, whereas trends were negative in the United States, Canada, parts of Africa, eastern Europe, Spain, Australia, and the former Soviet Union (*see Map 12*) (FAO 1999a:9–18; 51–53). Depending on the region, growth in harvests may stem from the exploitation of an underutilized resource, overexploitation of a fishery that will soon collapse, or enhancement of fisheries by stocking or introducing more productive species. The FAO found that, in every region, the major threat to fisheries is environmental degradation of freshwater habitat (FAO 1999a:19) (*see Box 3*).

Capacity of Freshwater Systems to Provide Food

Freshwater fish are extremely important in some regions for human nutrition and for local economies. At the global level, inland fisheries landings have been increasing since 1984. Most of this increase has occurred in Asia, Africa, and to a lesser degree in Latin America. In North America, Europe, Australia, and the former Soviet Union, landings have declined, whereas in parts of Oceania they have remained stable. Production of aquaculture has increased rapidly in the last decade, as has the yield of inland capture fisheries from introduced species or enhanced by stocking practices. However, successful stocking and introduction programs require functioning freshwater systems, which are being threatened by degradation (FAO 1999a:44).

In almost all regions of the world, the capacity of freshwater systems to support wild fish stocks is highly threatened because of habitat degradation and loss of fishery habitat (FAO 1999a:43). Overfishing and destructive fishing practices are also threatening stocks in parts of Asia and, on a more local level, in Africa and Latin America.

The increasing importance of recreational fishing, as well as the growing concern about biodiversity, pristine habitat, and other goods and services, such as clean water, is driving efforts in some countries to restore and rehabilitate freshwater systems.

Box 3

Threats and Issues Facing Inland Fisheries by Continent

Threats and issues in North America:

- Highly fragmented rivers (caused by dams, canals, and other projects).
- Non point source pollution.
- Increasing importance of recreational fisheries.
- Major fisheries dependent on stock enhancement.
- Increasing attention to river and habitat restoration.

Threats and issues in Central and South America:

- Water pollution from mining industry discharges and expanding urban centers.
- Localized overfishing.
- Increasing turbidity and sedimentation from agricul tural runoff and deforestation.
- Localized threats from dam construction.

Threats and issues in Europe:

- Highly fragmented rivers (caused by dams, canals, and other projects).
- Water pollution.
- Increasing importance of recreational fisheries.
- Some fisheries dependent on stock enhancement.
- Increasing attention to river and habitat restoration.

Threats and issues in Africa:

- Increasing habitat degradation.
- Localized water pollution and overfishing.
- Localized problems with species introductions.

Threats and issues in Asia:

- Habitat degradation.
- Interference of water flow by construction works.
- Overfishing and destructive fishing practices.
- High dependency on stock enhancement

Threats and issues in Oceania:

- Water pollution from pesticide and herbicide runoff.
- High sediment loads and rising salinity in inland waters.
- Increasing importance of recreational fisheries.

Sources: FAO 1999a; Arthington and Welcomme 1995; Remane 1997; FAO 1995b.

Inland Fisheries Information Status and Needs

The FAO database on inland fisheries is the most complete data set on fishery resources at the global level and it provides some useful insights on what is happening to freshwater fisheries worldwide. However it has several important limitations for assessing the capacity of freshwater systems to provide food.

First, the catch composition is not well known because reporting at the species level is very poor in most countries. Overall, nearly 45 percent of all inland capture fisheries are reported as "freshwater fish not elsewhere included (nei)" (FAO 1999a:4). This makes assessments of inland fisheries particularly difficult. Part of this lack of reporting at the species level is because of the limited representation in FAO statistics of the large diversity of freshwater fauna. Although there are 11,500 identified species of freshwater fish, the FAO lists only 100 of these species or species groups in its catch statistics (FAO 1999a:4). In Asia, the region with the largest inland fisheries production, 80 percent of the landings in 1992 were reported as "freshwater fish nei" (FAO 1995b:22).

Second, much of the inland fish catch comes from subsistence and recreational fisheries. Even though most of the catch from these fisheries is for human consumption, these landings are not reflected in the FAO catch statistics. Most national offices do not report on recreational fisheries and, because much of the subsistence catch is consumed locally, products do not always enter the market and therefore landings are not recorded (FAO 1995b:2, 3). This situation is so prevalent in many developing countries that recent evaluations carried out by the FAO show that actual catches are probably twice as large, and in some countries, three times as large, as the reported landings (FAO 1999a:4).

Finally, inland fisheries are usually dispersed over large areas, which makes data collection difficult and very expensive. National reporting offices, particularly in developing countries, are poorly funded and justifying expensive data collection for inland fisheries is becoming increasingly difficult, adding to the underreporting problem (FAO 1999a:4).

To improve our capacity to assess the condition of aquatic systems, catch statistics and trend analyses can be supplemented with national-level information on market demand, land-use change, and climatic factors. The FAO did this in its 1995 review, *The State of Inland Capture Fisheries*, and its fishery country profiles, which provide a more in-depth analysis of the capacity of freshwater systems in each region to sustain inland capture fish production.

There is an urgent need to improve the quality of the data on inland capture fisheries and those environmental and socioeconomic factors that affect their sustainability. Reporting should be at the species level, and ideally catch data should be collected and reported at the watershed or basin level (FAO 1999a:4). National-level statistics do not provide the necessary information to assess the condition of a particular body of water. This is especially true when a river or lake crosses international borders. Reporting at the watershed level could provide better insight into the relationship between upstream activities and downstream effects, and these could be incorporated into management plans.

Because the major threat to inland fisheries is environmental degradation, information on land-use change, water quality, water withdrawals, and species introductions is critical for assessing the condition and potential of a particular fishery. At present, these data are not collected at the watershed level, which makes their inclusion in analysis of fisheries resources difficult.

Finally, data collection on recreational fisheries and fishery enhancements, especially stocking and introduction programs, have to be incorporated in a systematic way into fisheries statistics. The consequences of enhancement practices should also be assessed more closely to ensure that the integrity of the ecosystem and its capacity to provide other goods and services is maintained.

BIODIVERSITY

Overview

One fundamental service provided by freshwater systems is habitat for a wide range of species. An estimated 12 percent of all animal species live in fresh water (Abramovitz 1996:7). Many others, including humans, depend on fresh water for their survival. In Europe, for example, 25 percent of birds and 11 percent of mammals use freshwater wetlands as their main breeding and feeding areas (EEA 1995:90). Although freshwater ecosystems have fewer species than marine and terrestrial habitats, species richness is very high when habitat extent is taken into account (*see Table 6*). There are 44,000 described aquatic species, which represent 2.4 percent of all known species, according to estimates from Reaka-Kudla (1997:90). Yet freshwater systems occupy only 0.8 percent of the Earth's surface (McAllister et al. 1997:5). In addition, scientists estimate that the number of known freshwater species is only a portion of the actual number of freshwater species. In fact, in the last 18 years, about 309 new freshwater species have been described each year (Nelson 1976, 1984, 1994).

Humans use these animals and plants for food, crops, skins, medicinal products, ornamental products (such as aquarium fish), biological control of insects and weeds, and increasingly for recreational purposes. In addition, freshwater biodiversity helps maintain ecosystem functions and ecosystem services, such as primary productivity, water purification, nutrient recycling, and waste assimilation.

Physical alteration, habitat loss and degradation, water withdrawal, overexploitation, pollution, and the introduction of nonnative species all contribute directly or indirectly to declines in freshwater species. These varied stresses affecting aquatic systems occur all over the world, although the particular effects of these stresses vary from watershed to watershed. In a recent study of freshwater fish by Harrison and Stiassny (1999), habitat alteration and the introduction of nonnative species were found to be the major causes driving the extinction of species. They study attributed 71 percent of extinctions to habitat alteration, 54 percent to the introduction of nonnative species, 29 percent to overfishing, 26 percent to pollution, and the rest to either hybridization, parasites and diseases, and intentional eradication (some extinctions may have had several causing factors, therefore, percentages do not add up to 100) (Harrison and Stiassny 1999:298–299).

Perhaps the best measure of the actual condition of freshwater biodiversity is the extent to which species are threatened with extinction. Globally, scientists estimate that more than 20 percent of the world's 10,000 described freshwater fish species have become extinct, threatened, or endangered in recent de-

Table 6

Species Richness by Ecosystem

Ecosystem	Habitat Extent	Percent Known Species*	Relative Species Richness**
Freshwater	0.8%	2.4%	3
Terrestrial	28.4%	77.5%	2.7
Marine	70.8%	14.7%	0.2

*Sum does not add to 100 percent because 5.3 percent of known symbiotic species are excluded.

**Calculated as the ratio between the percent species known and the percent area occupied by the ecosystem.

Source: McAllister et al. 1997.

cades (Moyle and Leidy 1992:140). This number, however, is considered to be a major underestimate (Bräutigam 1999:4). According to the 1996 IUCN *Red List of Threatened Animals*, 734 species of fish are classified as threatened, of which 84 percent are freshwater species (IUCN 1996:intro p. 37; McAllister et al. 1997:38). For some countries and regions, there is more detailed information on threatened status of aquatic species. In Europe, 42 percent of freshwater fish are threatened or endangered, in Iran, 22 percent, and in South Africa, 63 percent (Moyle and Leidy 1992:138). In the United States, which has comparatively detailed data on freshwater species, 37 percent of freshwater fish species, 67 percent of mussels, 51 percent of crayfish, and 40 percent of amphibians are threatened or have become extinct (Master et al.1998:6).

Assessing the condition of aquatic species is of critical importance if we are going to try and preserve the integrity of these ecosystems and the goods and services we derive from them. Unfortunately, data on freshwater species are not readily available for many countries and for most taxa. Given the limitation of the available data on freshwater biodiversity, we have attempted to provide indicators of biological importance (value) as well as indicators of condition of freshwater biodiversity. However, further research and data collection are urgently needed in this area.

Indicators of Biological Value

Freshwater species distributions are not well documented in many countries, particularly in developing regions of the world. Even in developed nations, data on freshwater biodiversity is available usually for the higher vertebrate species, but not for lower taxa. For instance, there is considerable information on fish, crustaceans, and molluscs for U.S. rivers and lakes, but much of the aquatic fauna and flora in Latin America is still unknown. Fish are the most studied group of freshwater species and, therefore, allow for a more in-depth analysis. For other groups, such as crustaceans or molluscs, it is possible to do only preliminary analyses based on expert opinion or regional analyses for those areas for which there is more data.

To assess the value of freshwater systems in terms of biological importance, we selected the following two indicators:

- Species richness and endemism at the global level.
- Biological distinctiveness at the regional level: the case of North America.

SPECIES RICHNESS AND ENDEMISM AT THE GLOBAL LEVEL

In an attempt to prioritize conservation areas for freshwater biodiversity, the World Wildlife Fund-US (WWF-US) and the World Conservation Monitoring Centre (WCMC) have carried out analyses to identify some of the most significant freshwater biodiversity areas of the world. WCMC produced a preliminary map of important areas for freshwater fish, crustacean, and mollusc diversity. These areas of high diversity focus on five groups of species: freshwater fish, molluscs, crabs, crayfish, and fairy shrimp, and were identified through expert opinion (Groombridge and Jenkins 1998). WCMC identified 136 areas of high freshwater biodiversity, 32 in Africa, 14 in Oceania, 45 in Eurasia, 18 in North America, and 27 in South America (*see Map 13*).

WWF-US has also identified areas of global importance for freshwater biodiversity based on species richness, species endemism, unique higher taxa (such as genera and families), unusual ecological or evolutionary phenomena, and global rarity of major habitat types (Olson and Dinerstein 1999). This analysis identified 53 freshwater ecoregions as outstanding areas for freshwater biodiversity. It is important to note that this ecoregional analysis, known as the Global 200, has also identified important terrestrial areas, such as grasslands and forests. Many of these selected forest and grasslands areas can also be considered "freshwater ecosystems" because they are seasonally flooded and represent important areas for freshwater as well as terrestrial biodiversity. Map 13 draws on these two efforts and presents a global view of important areas for freshwater biodiversity.

The 53 freshwater ecoregions highlighted by the Global 200 are classified into major habitat types that fall within one of the following systems: large rivers, larger river deltas, large river headwaters, small rivers, large lakes, small lakes, and xeric basins. These ecoregions vary in size and species composition. To illustrate the range of important freshwater ecoregions, we have selected two examples of these outstanding areas: the Amazon River with the surrounding flooded forests in Brazil, Peru, and Colombia, and the Chihuahuan basin in Mexico and the United States. The large ecoregion of the Amazon River and the surrounding flooded forests has an especially large concentration of freshwater and terrestrial biodiversity. The Amazon

River alone has more than 3,000 species of fish (Goulding 1985), with new species being identified at a fast rate. The flooded forests of this ecoregion are also rich in terrestrial and aquatic species, particularly those species that depend on these seasonally flooded forests as spawning and feeding habitat. On the other end of the spectrum with respect to size, the Chihuahuan xeric basin, which is composed of small springs and pools, is an outstanding center for evolutionary phenomena and endemism. More than half of the species found in the Chihuahuan basin are endemic, including 23 species of freshwater molluscs (Olson and Dinerstein 1999).

As Map 13 shows, the major concentrations of freshwater biodiversity (according to these two analyses) are in the tropical areas of Latin America, Africa, and Southeast Asia, as well as large parts of Australia, Madagascar, North America, the Yangtze basin in China, and the Amur basin in Russia. Because of the increase in available data for fish species, we have gone a step further and analyzed species richness and endemism by watershed for selected major watersheds of the world (*see Map 14*). The selected watersheds represent large basins that cross national borders and some smaller watersheds of regional significance. In all, these basins cover approximately 55 percent of the world's land area (excluding Antarctica).

Of the 108 watersheds analyzed, 27 have particularly high fish species richness. Of these, 56 percent are in the tropics, particularly Central Africa, mainland Southeast Asia, and South America, even though only about a third of all watersheds analyzed are tropical. High fish diversity is also found in central North America and in several basins in China and India. The pattern of unique species, or endemism, shows strong similarities to the pattern of species richness, particularly in Central Africa, South America, and Southeast Asia. In temperate regions, the Colorado, Rio Grande, and Alabama basins in North America stand out (for their size) as having large numbers of endemic fish.

Table 7 shows the three top ranking basins in terms of number of fish species and endemics by watershed size. Because there is a correlation between number of species and area, large watersheds tend to have more fish species than smaller ones (Oberdorff et al. 1995). To help eliminate bias in size, we classified basins into three categories. Large watersheds are those with an area equal to or greater than 1.5 million km^2, medium watersheds are those with areas between 400,000 and 1,499,999 km^2, and small watersheds have areas smaller than 400,000 km^2.

BIOLOGICAL DISTINCTIVENESS AT THE REGIONAL LEVEL: THE CASE OF NORTH AMERICA

At the regional level, WWF-US has also developed an index of biological distinctiveness (BDI) for each freshwater ecoregion in North America (Abell et al. 2000). The approach used is similar to the Global 200 ecoregion analysis, but at a different scale and with more detailed data. The freshwater ecoregions analyzed were delineated based on freshwater species distributions; in particular, the distributions of native freshwater fish, crayfish, and unionid mussels. These ecoregions generally coincide with watershed boundaries.

The BDI includes measures for species richness, species endemism, uniqueness of higher taxa, rarity of habitat type, and rarity of ecological or evolutionary phenomena (such as extraordinary salmon spawning runs). Species richness and endemism data were from a variety of published and unpublished sources on the distribution of over 2,200 North American species belonging to the following taxonomic groups: freshwater fish, crayfish, unionid mussels, amphibians, and aquatic and semiaquatic reptiles. The assessment of habitat type and ecological or evolutionary phenomena was based on expert opinion. Ecoregions were then classified into globally outstanding, continentally outstanding, bioregionally outstanding, or nationally important levels.

Table 7

Watersheds with High Fish Species Richness and Endemism

Watershed Size	Watershed	Number of Fish Species	Watershed	Number of Fish Endemics
Large Watersheds	Amazon	3,000	Amazon	1,800
	Congo	700	Congo	500
	Mississippi	375	Mississippi	107
Medium Watersheds	Rio Negro	600	Xi Jiang (Pearl River)	120
	Mekong	400	Orinoco	88
	Madeira	398	Paraguay	85
Small Watersheds	Lake Victoria	343	Lake Victoria	309
	Kapuas	320	Lake Tanganyika	216
	Lake Tanganyika	240	Salween	46

Source: Revenga et al. 1998.

Figure 8

Globally Outstanding Freshwater Ecoregions in North America

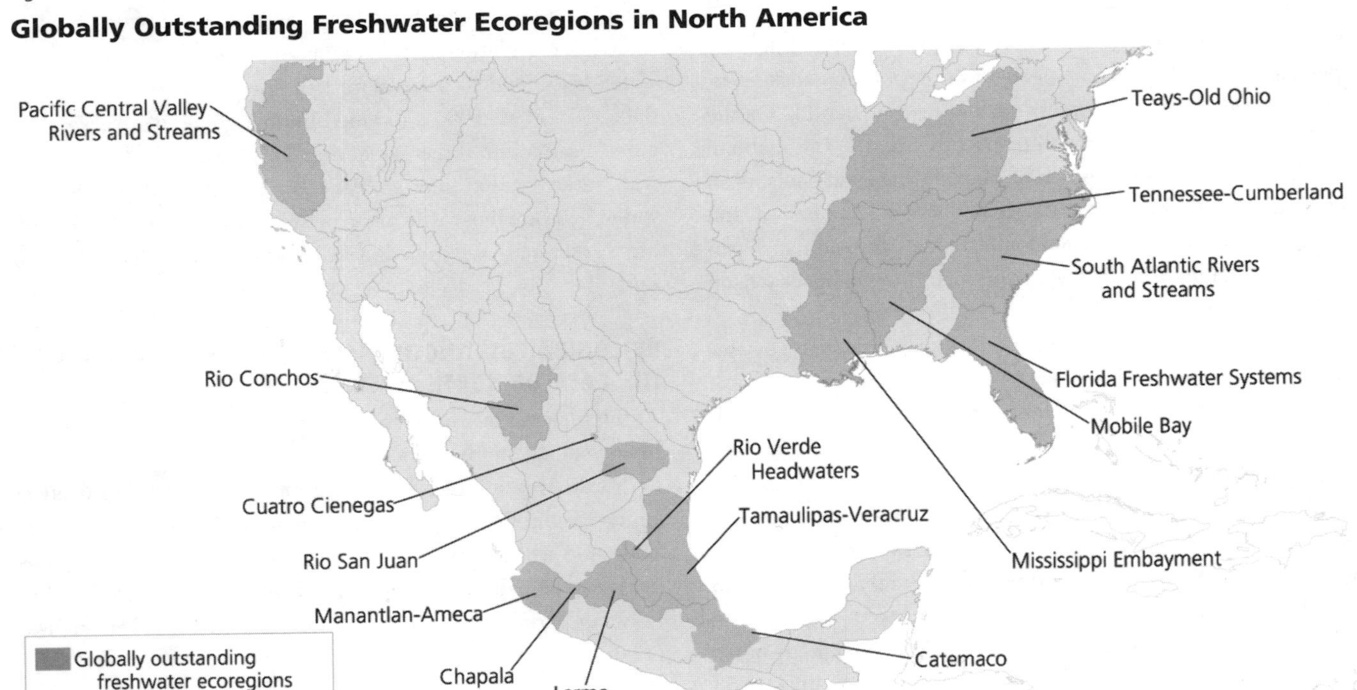

Source: Abell et al. 2000.

According to the WWF-US analysis, there are 16 globally outstanding ecoregions in North America, including the biologically rich and diverse ecoregions of Tennessee-Cumberland, Mobile Bay, and the Mississippi embayment. Globally outstanding ecoregions also include several freshwater systems in Mexico that have high degrees of endemism, such as the Lerma and the Cuatro Ciénegas ecoregions. Figure 8 shows those globally outstanding ecoregions in the United States and Mexico.

Condition Indicators of Biodiversity

One of the major challenges when assessing the condition of freshwater biodiversity is the lack of data on freshwater species. Many species remain unknown, particularly in developing countries, and monitoring of species populations is occurring only in localized areas, generally in developed countries. Despite this lack of data, we selected various indicators under the following categories to illustrate how freshwater systems are doing in terms of biodiversity:

Species population trends.

- Bird population trends in the United States and parts of Canada—wetland-dependent species.
- Global amphibian population census.

Threatened species.

- Imperiled fish and herpetofauna in North American freshwater ecoregions.
- Birds in Europe and the Middle East.

Presence of nonnative species.

- Introduced fish.
- Zebra mussel distribution in the United States.
- Global and U.S. distribution of water hyacinth.

SPECIES POPULATION TRENDS

Population trends represent some of the best indicators for measuring the condition of individual species and groups of species. When species with similar life histories and ecological traits are combined into groups and their population trends are analyzed, they provide additional insights into the overall health and condition of the habitat that they depend on. Continental- or global-level data sets on population trends for extended time periods are not readily available for many freshwater-dependent species. Therefore, we have selected two data sets to illustrate the condition of freshwater systems. These data sets are the North American Breeding Bird Survey (BBS) for wetland-dependent bird species in the United States and parts of Canada, and a global census on amphibian populations.

Bird population trends in the United States and parts of Canada

In 1966, the U.S. Department of the Interior launched the BBS. This survey is now being coordinated by the Patuxent Environmental Science Center in Maryland, and it covers the continental United States and parts of Canada. The survey is based on observations of breeding birds along more than 3,700 survey routes, each approximately 40 kilometers in length (BBS Website: http://www.mbr.nbs.gov/bbs). BBS data are extremely useful for monitoring the distribution and trends of particular species over time, and they provide one of the few such available continental-level data sets.

The Patuxent Environmental Science Center has grouped the more than 420 species of North American breeding birds into five breeding habitat groups: grassland, wetland-open water, successional-scrub, woodland, and urban species (Sauer et al. 1997). The wetland-open water group, discussed here, includes 86 species that depend on wetlands and open water, such as the sandhill crane, northern pintail, canvasback, and king rail. In terms of species richness, the BBS data show that for the period of 1982 to 1996, the largest numbers of wetland species (more than 60) are found in the Canadian provinces of Saskatchewan and Alberta, and in North Dakota and Montana in the United States. Similar numbers of species also occur locally in Florida, eastern Texas, southern Louisiana, northern California, and Minnesota. The fewest numbers of wetland species (less than five species), as is expected, are found in more arid areas along the Mexican-U.S. border from Texas to Arizona, as well as parts of southern Nevada, California, and Utah.

The population trend, covering the period from 1966 to 1998, reflects both increases and declines in the populations of wetland species across the United States and Canada. Wetland birds seem to be increasing from the Rocky Mountain region to the Pacific Northwest, the Great Lakes region, and in the central United States—especially in Texas— and from Louisiana north to Minnesota. Increases also have been recorded in areas of California and Manitoba. Declining trends prevail in eastern North America from Kentucky and Virginia south to Florida, parts of Quebec and Ontario, and the "prairie pothole" region—especially North Dakota. Significant declines are also noted in the southwestern states.

In the entire survey area, 41 percent of wetland species had a significant positive trend, while 15 percent had a significant negative trend. Among those species with negative population trends, the most significant declines were noted for the rusty blackbird, mottled duck, common tern, northern pintail, king rail, horned grebe, little blue heron, and herring gull (BBS Website: http://www.mbr.nbs.gov/bbs).

According to the BBS data, nearly 66 percent of wetland bird species have increasing population trends, the highest percentage of increasing species of any group of bird species (BBS Website: http://www.mbr.nbs.gov/bbs). However, the Patuxent Environmental Science Center recommends that these trend estimates be "viewed with considerable caution because wetland birds tend to be very poorly represented along most BBS routes." It points out that the preferred habitat of most of the wetland birds is not found along the BBS routes and that wetland species tend to be "very secretive and poorly censused by the BBS methodology" (BBS Website: http://www.mbr.nbs.gov/bbs).

Some of these species are monitored through other national surveys, such as waterfowl and other game species, which are surveyed by the Office of Migratory Bird Management of the U. S. Fish and Wildlife Service and the Canadian Wildlife Service. Other species groups, such as marsh-breeding birds, are poorly represented in the BBS and other existing monitoring programs because they tend to inhabit areas that are not readily accessible. According to the Marshbird Monitoring Program, "little is known about the abundance, population trends, or management needs of marsh bird species"; however, certain species such as the black and yellow rails and American bitterns are of concern because "they are thought to be rare or declining" (Marshbird Monitoring Website: http://www.mp1-pwrc.usgs.gov/marshbird/).

Global amphibian population census

Because of their combined terrestrial and aquatic lifecycle, diverse reproductive modes, and permeable skin, amphibians are good indicators of the health of ecosystems (Pelley 1998). They are more susceptible to climate change and pollution and usually act as "early warning systems" when changes in the environment are occurring.

Amphibian populations have declined or even become extinct in the past 50 years (DAPTF Website: http://www2.open.ac.uk/Ecology/J_Baker/JBtxt.htm). Many of these declines were due to habitat alteration, such as the drainage of wetlands, or to introduced predators and pollution. But over the past 20 years, scientists throughout the globe have documented dramatic declines or "atypical" fluctuations of amphibian populations. These declines have occurred even in apparently pristine habitats, such as national parks and nature reserves where human activity is very limited (Houlahan et al. 2000:752; Pelley 1998; Carey et al. 2000:1). The evidence of these declines is mostly anecdotal, mainly because of a lack of long-term population data. However, because of the rapid disappearance of amphibians in many parts of the world, IUCN's Species Survival Commission in 1991 launched the Declining Amphibian Populations Task Force (DAPTF). DAPTF is a network of more than 3,000 scientists working in 90 countries to "determine the nature, extent and causes of declines of amphibians throughout the world and promote means by which declines can be halted or reversed" (DAPTF Website: http://

www2.open.ac.uk/Ecology /J_Baker/JBtxt.htm). Map 15 presents the first results of DAPTF.

These results show that there have been marked declines in eastern Australia, southeast Brazil, Central America, and at higher altitudes in the United States and Canada. More complete data for Australia show that in the last 20 years, 13 percent of the 208 native amphibian species have been classified as threatened or vulnerable—with almost 8 species considered to be probably extinct (Tyler 1997). In Costa Rica, 20 of the 50 amphibian species that were found in a 30 km^2 study area of the Monteverde Cloud Forests Preserve have not been seen since 1990 (Pounds et al. 1997).

This global trend in amphibian populations is confirmed by another recently published study by Houlahan et al. (2000). The study analyzed 936 population data sets from 37 countries and concluded that amphibian populations in North America and western Europe declined by 15 percent each year for the 1960–66 period and 2 percent per year for the 1966–97 period. It should be noted that the sample size for the earlier years is relatively small (Houlahan et al. 2000:753). Because the data sets used in this study were mostly for northern latitudes, population declines for South America, Africa, and Australia were not assessed (Houlahan et al. 2000:753).

Although there is considerable uncertainty in determining whether these phenomena are caused entirely by anthropogenic changes in the environment alone, there is a general agreement among experts that population declines are caused by a combination of environmental factors acting synergistically. These factors include the following: an increased exposure to ultraviolet radiation resulting from the thinning of the stratospheric ozone layer; chemical pollution from pesticides, fertilizers, and herbicides; acid rain; pathogens; introduction of predators; and global climate change. Because of the wide range of factors that are possibly affecting amphibians worldwide, population declines have been interpreted as a symptom of the general environmental degradation (Carey et al. 2000; Lips 1998; Pelley 1998; DAPTF Website: http://www2.open.ac.uk/Ecology/J_Baker/JBtxt.htm).

THREATENED SPECIES

Species threat-status also represents a measure of the condition of a particular ecosystem. For example, if a large percentage of species in a particular river basin is highly threatened, it is likely that this risk is correlated to several factors affecting the condition of that particular basin. Information on the status of most freshwater species is not well known for many regions of the world, but there are some data sets that allow us to look at the condition of some of these groups. In particular, we have selected the status of fish and herpetofauna in theUnited States and the status of wetland-dependent birds in Europe and the Middle East. The indicator selection was based on data availability.

Imperiled fish and herpetofauna for North American freshwater ecoregions

Data on imperiled fish and herpetofauna for North America were aggregated into freshwater ecoregions by the WWF-US as part of its North America conservation assessment work on freshwater ecoregions (Abell et al. 2000). Imperiled species are those that are considered endangered, threatened, or vulnerable (Abell et al. 2000:75). The original data for the United States and Canada are from The Nature Conservancy's Natural Heritage database as of 1997. Additional fish data are from Williams et al. (1989). The data for Mexico came from the Comisión Nacional para el Conocimiento y Uso de la Biodiversidad (CONABIO)(1998) and Gonzales et al. (1995).

The analysis of the data in Map 16 show that in the majority of the western ecoregions more than 10 percent of fish species are imperiled. In 11 ecoregions more than 25 percent of the fish species are at risk, and in 3 ecoregions— Death Valley, Vegas-Virgin, and the Rio Verde headwaters —more than 50 percent of the fish species are classified as endangered, threatened, or vulnerable. The Vegas-Virgin ecoregion has the highest number of fish species at risk: 11 species, or 64 percent of its fish fauna.

For herpetofauna, the imperilment level shows a similar pattern as that of fish, with the western United States and Mexican ecoregions having more imperiled species (*see Map 16*). In three ecoregions, Death Valley, Vegas-Virgin, and Rio Lerma, more than 25 percent of the herpetofauna species are at risk. In Rio Lerma, 31 of 62 species are imperiled.

Birds in Europe and the Middle East

Birds can also be useful indicators for measuring the biological value and condition of freshwater systems. In Europe, 20 percent of regularly occurring bird species are dependent on inland wetlands. Of all bird species with "unfavorable conservation status" in Europe—those that are categorized as endangered, vulnerable, rare, or declining—30 percent are inland wetland-dependent species (Tucker and Evans 1997:133). Inland wetlands in Europe are also the major habitat for eight globally threatened species of birds, including the slender-billed curlew—one of the world's most threatened bird species and classified as critically endangered by IUCN (Collar et al. 1994:85). The major threat to these populations throughout Europe is loss of wetland habitat.

In the Middle East, BirdLife International has identified 391 Important Bird Areas (IBA). IBAs include areas that support important concentrations of bird species globally or regionally, areas with important threatened bird species, or sites that represent unique restricted habitats of importance to birds (Evans

1994:20–27, 31). Wetlands are the main habitat type for the IBAs in the Middle East. Half of the 391 IBAs are wetlands, a third of which are coastal wetlands and the rest classified as nonmarine wetlands (Evans 1994:31). Even though wetlands are a critical habitat for bird species in this region, 78 percent or 153 IBAs with nonmarine wetlands as their principal habitat are unprotected and only 6 percent have a high degree of legal protection in the region. Wetland IBAs are also the most threatened, with 70 sites classified as moderately to highly threatened, 63 sites classified as having low threat, and 24 sites for which the threat has not been quantified (BirdLife International 1999). The major threat to these areas is wetland loss or modification.

In Europe, BirdLife International has identified 3,619 IBAs in 51 countries that occupy an area of 931,700 km^2, of which 69 percent contain wetland habitats. Of these IBAs with wetlands, 57 percent have standing freshwater habitats, such as lakes, 44 percent have riparian habitat surrounding rivers and streams, and 36 percent have a combination of fen, mire, bog, and spring habitats (some IBAs may contain two or more of these habitat types) (Heath and Evans 2000:43). Overall, 544 IBAs in this region are threatened by drainage, 157 of which are highly threatened (Heath and Evans 2000:56).

PRESENCE OF NONNATIVE SPECIES

The introduction of nonnative species is the second-leading cause, after habitat degradation, of species extinction in freshwater systems (Hill et al. 1997:1). Exotic species affect native faunas through predation, competition, disruption of food webs, and the introduction of diseases. The spread of exotic species is a global phenomenon, one that is increasing with the spread of aquaculture, shipping, and global commerce. Introduced species, both intentional and accidental, include a variety of species groups from fish and higher plants (such as water hyacinth) to invertebrates and microscopic plants (such as, dinoflagellates). Worldwide, twothirds of the freshwater species introduced into the tropics and more than 50 percent of those introduced to temperate regions have become established (Welcomme 1988:29).

Species introduction can have serious economic and ecological consequences for freshwater ecosystems. While the effect of nonnative species is worldwide in scope, the majority of the information about them is largely anecdotal or comes from studies and data collection on species introduced to developed countries, such as the United States. Because comprehensive data on nonnative species and their effects on biodiversity and ecosystem condition are not available at the global or regional level, we have selected the following cases as indicators of freshwater ecosystem condition: introduced fish, the zebra mussel, and the water hyacinth.

Introduced fish

Nonnative fish introductions are common in most parts of the world, and they are an increasingly important component of aquaculture (FAO 1999a:25–26). Introductions are usually done to enhance food production and recreational fisheries, or to control pests such as mosquitoes and aquatic weeds. Introduced fish, for example, account for 96.2 percent of fish production in South America and 84.7 percent in Oceania (Garibaldi and Bartley 1998).

The introduction of nonnative fish, however, has its ecological costs. A survey of 31 studies of fish introductions in Europe, North America, Australia, and New Zealand found that, in 77 percent of the cases, native fish populations were reduced or eliminated following the introduction of nonnative fish. In 69 percent of the cases, the decline followed the introduction of a single fish species, with salmonids responsible for the decline of native species in half of these cases (Ross 1991:363). In North America, there have been recorded extinctions of 27 species and 13 subspecies of fish in the past 100 years. The introduction of alien species was found to be a contributing factor in 68 percent of these extinctions, although in almost every case there were multiple stresses contributing to each extinction, such as habitat alteration, chemical pollution, hybridization, and overharvesting (Miller et al. 1989:22).

Accidental introductions can also have devastating effects on indigenous fish and the fisheries they support, as happened with the sea lamprey (*Petromyzon marinus*) in the Great Lakes of the United States. This fish first appeared in Lake Ontario in 1835 and by the 1940s had extended its range into Lake Superior. Preying on native freshwater fish, the lamprey has been a contributing factor in the decline of several species, including walleye, lake trout, ciscoes, and introduced salmon. Its presence was a factor in the crash of the lake trout fishery in Huron, Michigan, and Superior lakes in the 1940s and 1950s. Sea lamprey predation, combined with overfishing and hybridization, led to the extinction of three ciscoe species in this century. In 1991, efforts to control sea lampreys through chemical and mechanical means cost Canada and the United States US$8 million, with an additional US$12 million spent on lake trout restoration (Fuller et al. 1999:19–21).

There are distinct trade-offs between the value of introducing fish for human consumption and the negative impact on native aquatic ecosystems. Two species of tilapia (Mozambique tilapia and Nile tilapia), the common carp, and two species of trout (rainbow and brook trout) have established self-sustaining populations around the world and contribute significantly to food production (FAO 1999a:27). In 1997, for example, 2.2 million metric tons of common carp and 742,000 metric tons of Nile tilapia were produced through aquaculture worldwide (FAO 1998). Other species that have been introduced worldwide for mosquito and weed control include the mosquito fish, guppy, and grass carp.

Map 17 shows the degree of international introductions by country as recorded in the FAO Database on Introductions of Aquatic Species (DIAS). The largest percentage of these introductions—35.5 percent—took place between 1940 and 1979 (DIAS Website: http://www.fao.org/fi/statist/fisoft/dias/index.htm). It should be noted that the DIAS database considers only species introduced from one country to another and not within-country introductions or translocations.

Introduced species account for almost 10 percent of global aquaculture production (DIAS Website: http://www.fao.org/fi/statist/fisoft/dias/index.htm), but they have largely unquantified negative effects on many native species in the systems to which they are introduced. By feeding at the bottom of lakes and rivers, carp, for example, increase siltation and turbidity, causing decreases in water clarity and negative impacts on native species (Fuller et al. 1999:69). They have been associated with the disappearance of native fish in Argentina, Venezuela, Mexico, Kenya, India, and elsewhere (Welcomme 1988). These fish also have been introduced into areas where they are not used for aquaculture, resulting in dramatic changes in fish community structures in many places. In the Salton Sea of California, Mozambique tilapia are considered a major factor in the decline of the desert pupfish (Fuller et al. 1999:440). In Australia, exotic fish species are the leading cause in the decline of 22 species of native fish that are classified as endangered, vulnerable, or rare (Wager and Jackson 1993).

Zebra mussel distribution in the United States

The invasion and spread of the zebra mussel (*Dreissena polymorpha*) in the United States is one of the few long-term documented cases that can illustrate the cost to biodiversity and to other goods and services derived from rivers and lakes. Native to eastern Europe, the zebra mussel had spread to all major European rivers by the early 19th century. The mussel was first discovered in the Great Lakes in 1988, and expanded its range in North America rapidly in the following decade (USGS zebra mussel website: http://nas.er. usgs.gov/zebra.mussel/docs/sp_account.html#HDR1). The mussels were introduced through the release of ballast water from transatlantic cargo ships. Attaching themselves to ships, the mussels within a few years spread throughout the Great Lakes and into the Mississippi River drainage area (USGS zebra mussel website: http://nas.er.usgs.gov/zebra.mussel/docs/sp_account.html#HDR1). Map 18 shows the rapid expansion of the zebra mussel from the Great Lakes to other rivers and lakes in the eastern part of the United States.

Zebra mussels are small and attach themselves to many man-made underwater structures, such as intake pipes for power plants, locks, and dams, and other industrial infrastructure. Utilities, industrial facilities, and drinking water treatment plants have been particularly affected. The economic costs of eradicating the mussels are enormous. Survey data from over 400 facilities in the United States and Canada found that the cost of zebra mussel eradication from 1989 to 1995 totaled over US$69 million, with annual costs rising from US$234,140 in 1989 to US$17,751,000 in 1995 (O'Neill 1996:1, 2). These figures represent gross underestimates of the total costs because they are based on a small sample of the total number of potentially affected facilities. An estimated projection of total costs from 1989 to 1995 is in the range of US$300 million to US$400 million (O'Neill, personal communication, 1999).

Introduced zebra mussels also have negative ecological effects. A study of zebra mussels in western Lake Erie found that zebra mussel infestation from 1989 to 1991 caused the deaths of almost all native clams at 17 sampling stations, leading to a collapse of clam biodiversity and the near-extinction of many native clam species. No living native clam species were found at the sampling stations by 1991 (Schloesser and Nalepa 1994:2238–2239). In the St. Croix River, a federally designated wild and scenic river in the upper Mississippi River basin, the only known viable population of the winged mapleleaf clam (*Quadrula frugosa*) is currently threatened by the advancing zebra mussels (USGS zebra mussel Website: http://nas.er.usgs.gov/zebra.mussel/docs/ sp_account.html#HDR1).

Global and U.S. distribution of water hyacinth

Water hyacinth (*Eichhornia crassipes*) is another example of a widespread exotic species that is causing considerable economic and ecological damage in many parts of the world. This plant, thought to be indigenous to the upper reaches of the Amazon basin, was spread throughout much of the planet for use as an ornamental beginning in the mid-19th century (Gopal 1987:1). By 1900 it had spread to every continent except Europe and has now achieved pan-tropical distribution. The plant spreads quickly to new rivers and lakes in the tropics, clogging waterways and infrastructure, reducing light and oxygen in freshwater systems, and causing changes in water chemistry and species assemblages (Hill et al. 1997). These changes can disrupt the livelihoods of local communities that depend on goods and services derived from these systems (Hill et al. 1997). For instance, in Papua New Guinea, water hyacinth has been confirmed in almost 100 locations with the most serious infestation reported for the Sepik River, where it has had "devastating effects on socioeconomic structure and on the environment" (Harley et al. 1997). The spread of hyacinth impairs fishing activities, travel to market to sell and buy produce, and access to fishing areas. In addition, hyacinth and other aquatic weeds act as vectors in the life cycles of insects that transmit diseases, such as schistosomiasis and lymphatic filariasis (Bos 1997). Even in its native range of South America, water hyacinth has become a pest in Guyana and Surinam and has spread into adjacent regions, creating problems in rivers and reservoirs by clogging dams and intake valves in Argentina, Bolivia, Cuba, and Mexico (Harley et al. 1997).

Unfortunately, despite the many problems associated with the water hyacinth, thorough studies quantifying its distribution, its impact on biodiversity, and its socioeconomic parameters within freshwater systems and rural riparian communities are lacking. The United States has a comparably comprehensive data set on the distribution of this alien species. In the United States, the aquatic plant has established populations in most of Florida, Louisiana, and along the coastal states of the Gulf of Mexico, and into basins in eastern Texas. The southern coastline of California and the Sacramento River basin have also been affected by the spread of the hyacinth (USGS invasive species Website: http://nas.er.usgs.gov/plants/maps/smec.gif). Map 18 shows the distribution of water hyacinth across the United States.

Water hyacinth has also spread to most rivers and lakes in the tropics. A 1987 study by B. Gopal found that the plant had spread to most regions of tropical and subtropical Asia, Africa, and Central America (Gopal 1987). This distribution pattern was confirmed by a more recent review by Harley et al. (1997). In recent years, water hyacinth has spread very rapidly in Africa, from the Nile Delta and the Congo basin to regions in West Africa (particularly Cote d'Ivoire, Benin, and Nigeria), the equatorial zone of East Africa, and to the southern part of the continent, specifically Zambia, Zimbabwe, Mozambique, and South Africa (Harley et al. 1997). In Australia, water hyacinth has spread along the entire coast, although its impact is being reduced by biological control. It has also spread to many Pacific Islands, including New Zealand. In South America, water hyacinth is found in the majority of the Amazon tributaries, as well as in Guyana, Surinam, Argentina, Paraguay, Uruguay, the northern part of Colombia, Venezuela, and the central rivers in Chile. It is also found in most of Central America and the Caribbean islands (Gopal 1987:46–60; Harley et al. 1997). In Asia, it has widely spread through the Southeast and India. In Europe, however, it has been reported only in Portugal (Gopal 1987:46–60).

Capacity of Freshwater Systems to Sustain Biodiversity

Habitat degradation, physical alteration through dams and canals, water withdrawals, overharvesting, pollution, and the introduction of nonnative species have all taken a heavy toll on freshwater biodiversity. As a consequence, the capacity of freshwater ecosystems to support biodiversity is highly degraded at a global level. Because of the lack of data for most freshwater species and the difficulty quantifying the impacts of some pressures, such as introduction of nonnative species, it is hard to provide a quantifiable measure of the condition of freshwater biodiversity around the world. However, studies done for North American freshwater fauna, for which there are more data, show that species are being lost at an "ever-accelerating rate" (Moyle and Leidy 1992:163) and suggest that future extinction rates are five times higher for freshwater animal species than for terrestrial ones (Ricciardi and Rasmussen 1999:1221). Indeed, of all the ecosystems assessed under PAGE, freshwater ecosystems are in by far the worst condition from the standpoint of their ability to support biological diversity. More than 20 percent of the world's 10,000 freshwater fish species have become extinct, threatened, or endangered in recent decades, and many more aquatic-dependent species, such as mussels, birds, and plants are also highly threatened. Rising global demand for water and food will increase the already considerable pressures on freshwater ecosystems, putting more of the species and ecosystem processes at risk.

Biodiversity Information Status and Needs

Direct measurements of the condition of biodiversity in freshwater systems are sparse worldwide. As mentioned throughout this section, global data on freshwater biodiversity are lacking for many developed nations and most of the developing world. This makes analyzing population trends impossible or limited to a handful of well-known species.

There is excellent trend data for bird populations in the United States and Canada. For other regions of the world, however, databases on the distribution of Important Bird Areas such as those from BirdLife International, are of high quality but lack long-term population trends. Information on nonnative species is frequently anecdotal and often limited to records of the presence or absence of a particular species, without documentation of the effects on the native fauna and flora. Regional and local spatial distribution on certain invasive species is available for few species, mostly in the United States and Australia.

Some of these data gaps are being filled through studies that draw on local experts, museum collections, and field inventories. However, more information on species assemblages, interactions, and population trends is urgently needed to assess the condition of freshwater systems. At the global level, many countries probably cannot mount this data collection effort because of the high monitoring costs. At a minimum, there should be monitoring of key indicator species, as well as monitoring of the presence or introduction of nonnative species and their impacts on native fauna and flora. The development of a simple and more affordable IBI at the watershed level may be a reasonable approach for many countries.

Bibliography

Abell, R.A., D.M. Olson, E. Dinerstein, P.T. Hurley, J.T. Diggs, W. Eichbaum, S. Walters, W. Wettengel, T. Allnutt, C.J. Loucks, and P. Hedao. 2000. *Freshwater Ecoregions of North America: A Conservation Assessment*, Washington, DC: World Wildlife Fund-United States.

Abramovitz, J. N. 1996. *Imperiled Waters, Impoverished Future: The Decline of Freshwater Ecosystems.* Worldwatch Paper 128. Washington, DC: Worldwatch Insititute.

Achieng, A. P. 1990. "The Impact of the Introduction of Nile Perch, *Lates niloticus (L.)*, on the Fisheries of Lake Victoria." *Journal of Fish Biology* 37(Supplement A): 17–23.

Alcamo, J., T. Henrichs, and T. Rösch. 2000. *World Water in 2025 — Global Modeling and Scenario Analysis for the World Commission on Water for the 21st Century.* Report A0002. Kassel, Germany: Center for Environmental Systems Research, University of Kassel.

Arthington, A. H. and R. L. Welcomme. 1995. "The Condition of Large River Systems of the World," pp. 44–75 in *Condition of the World's Aquatic Habitats.* Proceedings of the World Fisheries Congress, Theme 1. N. B. Armantrout and R. J. Wolotira, Jr., eds. Lebanon, New Hampshire, U.S.A.: Science Publishers Inc.

Avakyan, A. B. and V. B. Iakovleva. 1998. "Status of Global Reservoirs: The Positionin the Late Twentieth Century." *Lakes & Reservoirs: Research and Management* 3: 45–52.

Bacalbaça-Dobrovici, N. 1989. "The Danube River and its Fisheries," pp. 455–468 in *Proceedings of the International Large River Symposium.* D. P. Dodge, ed. Canadian Special Publication of Fisheries and Aquatic Science 106. Ottawa, Canada: Department of Fisheries and Oceans.

Barbier, E. B. and J. R. Thompson. 1998. "The Value of Water: Floodplain versus Large-Scale Irrigation Benefits in Northern Nigeria." *Ambio* 27 (6): 434–440.

Barbour, M. T., J. Gerritsen, and J. S. White. 1996. *Development of the Stream Condition Index (SCI) for Florida.* Florida Department of Environmental Protection. Owings Mills, Maryland, U.S.A.: Tetra Tech Inc.

BirdLife International. 1999. Georeferenced data on threat status for wetlands in the Middle East provided to WRI for the PAGE analysis.

Bos, R. 1997. "The Human Health Impact of Aquatic Weeds," in E.S. Delfosse and N.R. Spencer, eds., *Proceedings of the International Water Hyacinth Consortium.* Washington, DC: World Bank. Available on-line at: http://www.sidney.ars.usda.gov/scientists/nspencer/water_h/appendix4.htm

Brady, S. J. and C. H. Flather. 1994. "Changes in Wetlands on Nonfederal Rural Land of the Conterminous United States from 1982 to 1987." *Environmental Management* 18 (5): 693–705.

Bräutigam, A. 1999. "The Freshwater Crisis." *World Conservation*, 30 (2): 4-5.

Breeding Bird Survey Website. Available on-line at: http://www.mbr.nbs.gov/bbs. Viewed 7/31/00.

Briscoe, J. 1999. "Water Resources Management in Yemen – Results of a Consultation." Office memorandum. Washington, DC: The World Bank.

British Geological Survey. 1996. *Characterisation and Assessment of Groundwater Quality Concerns in Asia-Pacific Region: The Aquifers of the Asia-Pacific Region–An Invaluable but Fragile Resource.* UNEP/DEIA/AR.96-1. Prepared on the behalf of the U.N. Environment Programme (UNEP) and the World Health Organization. Nairobi, Kenya: UNEP.

Carey, C., H. R. Heyer, J. Wilkinson, R. A. Alford, J. W. Arntzen, T. Halliday, L. Hungerford, K. R. Lips, E. M. Middleton, S. A. Orchard, and A. S. Rand. 2000. "Amphibian Declines and Environmental Change: An Overview." Draft manuscript. Boulder, Colorado, U.S.A.: University of Colorado.

Carlson, C. A. and R. T. Muth. 1989. "The Colorado River: Lifeline of the American Southwest," pp. 220–239 in *Proceedings of the International Large River Symposium.* D. P. Dodge, ed. Canadian Special Publication of Fisheries and Aquatic Science 106. Ottawa, Canada: Department of Fisheries and Oceans.

CEQ (Council on Environmental Quality). 1995. *Enviromental Quality — Twenty-Fith Anniversary Rerpot.* Washington, DC: The Council for Environmental Quality.

Chenje, M. and P. Johnson, eds. 1996. *Water in Southern Africa.* A report by the Southern African Development Community, IUCN-The World Conservation Union, and Southern African Research & Documentation Centre. Harare, Zimbabwe: Print Holdings.

CIESIN (Center for International Earth Science Information Network), International Food Policy Research Institute, and World Resources Institute. 2000. *Gridded Population of the World, Version 2.* Palisades, New York: CIESIN and Columbia University. Available on-line at: http://sedac.ciesin.org/plue/gpw.

Collar, N. J., M. J. Crosby and A. J. Stattersfield. 1994. *Birds to Watch 2: the World List of Threatened Birds.* No.4, BirdLife Conservation Series, Duncan Brooks, ed. Cambridge, U.K.: BirdLife International.

CONABIO (Comisión Nacional para el Conocimiento y Uso de la Biodiversidad). 1998. *Programa de Cuencas Hidrológicas Prioritarias y Biodiversidad de México de la Comisión Nacional para el Conocimiento y Uso de la Biodiversidad.* Primer Informe Técnico. México D.F.: CONABIO/USAID/FMCN/WWF.

Cosgrove, W. J. and F. R. Rijsberman. 2000. *World Water Vision: Making Water Everybody's Business.* Report prepared for the World Water Council. London, U. K.: Earthscan Publications Ltd.

Dahl, T. E. 1990. *Wetland Losses in the United States 1780s to 1980s.* Washington, DC: U.S. Department of the Interior, Fish and Wildlife Service.

Dahl, T. E. and C. E. Johnson. 1991. *Status and Trends of Wetlands in the Conterminous United States, Mid-1970's to Mid-1980's.* Washington, DC: U.S. Department of the Interior, Fish and Wildlife Service.

DAPTF (Declining Amphibian Populations Task Force) Website available on-line at: http://www2.open.ac.uk/Ecology/J_Baker/JBtxt.htm. Viewed 8/19/99.

Darras, S., M. Michou, and C. Sarrat. 1999. *A First Step Toward Identifying a Global Delineation of Wetlands.* IGBP-DIS Wetlands Data Initiative. IGBP-DIS Working Paper No. 19. Toulouse, France: IGBP-DIS.

Davies, S.P., L. Tsomides, D.L. Courtemanch, and F. Drumond. 1995. *Maine Biological Monitoring and Biocriteria Development Program.* Augusta, Maine, U.S.A.: Maine Department of Environmental Protection.

DeShon, J.E. 1995. "Development and Application of the Invertebrate Community Index (ICI)," pp. 217–243 (Chapter 15) in *Biological Assessment and Criteria: Tools for Water Resource Planning and Decision Making,* W.S. Davis and T. Simon, eds. Boca Raton, Florida, U.S.A.: Lewis Publishers.

DHI Water and Environment. 1999. Maps provided to WRI for the PAGE analysis. Hørsholm, Denmark.

DIAS (Database on Introduction of Aquatic Species) Website. 2000. Available on-line at: http://www.fao.org/fi/statist/fisoft/dias/index.htm. Viewed 9/11/00.

Dugan, P. J. and T. Jones. 1993. "Ecological Change in Wetlands: A Global Overview," pp. 34–38 in *Waterfowl and Wetland Conservation in the 1990s: A Global Perspective.* Proceedings of an IWRB Symposium. St. Petersburg Beach, Florida, U.S.A. 12–19 November 1992. M. Moser, R.C. Prentice, and J. van Vessem, eds. IWRB Special Publication No. 26. Slimbridge, U. K.: The International Waterfowl and Wetlands Research Bureau (IWRB).

Dynesius, M. and C. Nilsson. 1994. "Fragmentation and Flow Regulation of River Systems in the Northern Third of the World." *Science* 266: 753–762.

EDC (U.S. Geological Survey's EROS Data Center). 1999. HYDRO 1K Dataset available on-line at: http://edcdaac.usgs.gov/gtopo30/hydro/. Sioux Falls, South Dakota, U.S.A.: USGS EDC and United Nations Environment Programme/Global Resource Information Database (UNEP/GRID).

EEA (European Environment Agency). 1994. *European Rivers and Lakes: Assessment of Their Environmental State.* P. Kristensen and H. O. Hansen, eds., EEA Environmental Monographs 1, National Environmental Research Institute, Copenhagen, Denmark: Danish Ministry of Environment and Energy.

EEA (European Environment Agency).1995. *Europe's Environment: The Dobříš Assessment.* D.Stanners and P. Bourdeau, eds. Copenhagen, Denmark: European Environment Agency.

EEA (European Environment Agency). 1998. *Europe's Environment: The Second Assessment.* Copenhagen, Denmark: European Environment Agency.

EEA (European Environment Agency). 1999. *Environment in the European Union at the Turn of the Century.* Environmental Assessment Report No. 2. Copenhagen, Denmark: European Environment Agency.

Emteryd, O., D.Q. Lu, and N. Nykvist. 1998. "Nitrate in Soil Profiles and Nitrate Pollution of Drinking Water in the Loess Region of China." *Ambio* 27 (6): 441–443.

EPA (Environmental Protection Agency). 1999. U.S. EPA Office of Water Website available on-line at: http://www.epa.gov/iwi/ Viewed 7/25/00.

ESRI (Environmental Systems Research Institute). 1992. *ArcUSA 1:214 GIS Database CD-Rom.* Redlands, California, U.S.A.: ESRI.

ESRI (Environmental Systems Research Institute). 1996. "World Countries 1995," in *ESRI Data and Maps. Volume 1. CD-Rom.* Redlands, California, U.S.A.: ESRI.

Evans, M. I. 1994. *Important Bird Areas in the Middle East.* No.2, BirdLife Conservation Series, Duncan Brooks, ed., Cambridge, U.K.: BirdLife International.

Falkenmark, M. and G. Lindh. 1993. "Water and Economic Development," pp. 80–91 in P. Gleick, ed., *Water in Crisis: A Guide to the World's Fresh Water Resources.* Oxford, U.K.: Oxford University Press.

Falkenmark, M. and C. Widstrand. 1992. "Population and Water Resources: A Delicate Balance." *Population Bulletin* 47 (3): 1–36.

FAO (Food and Agriculture Organization of the United Nations). 1995a. *Effects of Riverine Inputs on Coastal Ecosystems and Fisheries Resources.* FAO Fisheries Technical Paper No. 349. Rome, Italy: FAO.

FAO (Food and Agriculture Organization of the United Nations), 1995b. *Review of the State of World Fishery Resources: Inland Capture Fisheries.* FAO Inland Water Resources and Aquaculture Service, Fishery Resources Division.FAO Fisheries Circular No. 885. Rome, Italy: FAO.

FAO (Food and Agriculture Organization of the United Nations). 1998. *Global Production Dataset 1984–1997, Global Capture Dataset 1984–1997,* and *Aquaculture Quantities Dataset 1984–1997,* Fishery Statistics Databases, downloadable with Fishstat-Plus software at: http://www.fao.org/WAICENT/ FAOINFO/FISHERY/statist/FISOFT/FISHPLUS.HTM.FISHSTAT PLUS, Version 2.19 by Yury Shatz. Rome, Italy: FAO.

FAO (Food and Agriculture Organization of the United Nations). 1996. "Malawi Fishery Country Profile." FAO Fisheries Department Website available on-line at: http://www.fao.org/fi/fcp/malawie.htm. Viewed 6/27/00.

FAO (Food and Agriculture Organization of the United Nations). 1999a. *Review of the State of World Fishery Resources: Inland Fisheries*. FAO Inland Water Resources and Aquaculture Service, Fishery Resources Division, FAO Fisheries Circular No. 942. Rome, Italy: FAO.

FAO (Food and Agriculture Organization of the United Nations). 1999b. *The State of World Fisheries and Aquaculture 1998*. Rome, Italy: FAO Fisheries Department.

FAO (Food and Agriculture Organization of the United Nations). 1999c. Data on location of dams for Africa provided to WRI for PAGE Analysis. Rome, Italy: FAO.

Fekete, B., C. J. Vörösmarty, and W. Grabs. 1999. *Global, Composite Runoff Fields Based on Observed River Discharge and Simulated Water Balance*. World Meteorological Organization Global Runoff Data Center Report No. 22. Koblenz, Germany: WMO-GRDC.

Finlayson, C.M. and N.C. Davidson. 1999. *Global Review of Wetland Resources and Priorities for Wetland Inventory: Summary Report.* Wageningen, Netherlands: Wetlands International and Jabiru, Australia: the Environmental Research Institute of the Supervising Scientists.

Foster, S., A. Lawrence, and B. Morris. 1998. *Groundwater in Urban Development: Assessing Management Needs and Formulating Policy Strategies.* World Bank Technical Paper No. 390. Washington, DC: The World Bank.

Foster, S., J. Chilton, M. Moench, F. Cardy, and M. Schiffler. 2000. *Groundwater in Rural Development: Facing the Challenges of Supply and Resource Sustainability.* World Bank Technical Paper No. 463. Washington, DC: The World Bank.

Foster, S. British Geological Survey. 2000. Personal Communication. 5 July.

Frolking, S., X. Xiao, Y. Zhuang, W. Salas, and C. Li. 1999. "Agricultiral Land-Use in China: A Comparison of Area Estimates from Ground-Based Census and Satellite-Borne Remote Sensing." *Global Ecology & Biogeography* 8 (5): 407–416.

Fuller, P.L., L.G. Nico, and J.D. Williams. 1999. *Nonindigenous Fishes Introduced into Inland Waters of the United States*. American Fisheries Society, Special Publication no.27. Bethesda, Maryland, U.S.A.: American Fisheries Society.

Ganasan, V. and R. M. Hughes. 1998. "Application of an Index of Bbiological Integrity (IBI) to Fish Assemblages of the Rivers Khan and Kshipra (Madhya Pradesh), India." *Freshwater Biology* 40: 367–383.

Garibaldi, L. and D. M. Bartley, 1998. The database on introductions of aquatic species (DIAS). Food and Agriculture Organization of the United Nations (FAO) Aquaculture Newsletter (FAN), no. 20. Available on the Web at http://www.fao.org/fi/statist/fisoft/dias/index.htm.

GLCCD (Global Land Cover Characteristics Database), Version 1.2. 1998. Loveland, T.R., B.C. Reed, J.F. Brown, D.O. Ohlen, Z. Zhu, L. Yang, and J. Merchant. 2000. "Development of a Global Land Cover Characteristics Database and IGBP DISCover from 1-km AVHRR data." *International Journal of Remote Sensing* 21 (6-7): 1303–1330. Data available on-line at: http://edcdaac.usgs.gov/ glcc/glcc.html.

Gleick, P. H.1998. *The World's Water 1998-1999: The Biennial Report on Freshwater Resources.* Washington, DC: Island Press.

Gonzales, A., J. L. Camarillo, F. Mendoza, and M. Mancilla. 1995. "Impacts of Expanding Human Populations on the Herpetofauna of the Valley of Mexico." *Herpetological Review* 17: 30–31.

Gopal, B. 1987. *Water Hyacinth.* Aquatic Plant Studies 1, Amsterdam and New York: Elsevier Science.

Goulding, M. 1985. "Forest Fishes of the Amazon," in *Key Environments: Amazonia,* G.T. Prance and T.E. Lovejoy, eds., Oxford, U.K.: Pergamon Press.

Groombridge B. and M. Jenkins. 1998. *Freshwater Biodiversity: a Preliminary Global Assessment.* World Conservation Monitoring Centre, Cambridge, U.K.: World Conservation Press.

Harley, K., M.H. Julien, and A.D. Wright. 1997. "Water Hyacinth: a Tropical Worldwide Problem and Methods for its Control," in E.S. Delfosse and N.R. Spencer, eds., *Proceedings of the International Water Hyacinth Consortium.* Washington, DC: World Bank. Available on-line at: http://www.sidney.ars.usda.gov/scientists/ nspencer/water_h/ appendix7.htm

Harrison, I. J. and M. J. Stiassny. 1999. "The Quiet Crisis: A Prelimminary Listing of the Freshwater Fishes of the World that Are Extinct or 'Missing in Action'," pp. 271–331 in *Extinctions in Near Time,* MacPhee, ed. New York, New York: Kluwer Academic/Plenum Publishers.

Heath, M. F. and M. I. Evans, eds. 2000. *Important Bird Areas in Europe: Priority Sites for Conservation.* 2 vols. BirdLife Conservation Series No. 8. Cambridge, U. K.: BirdLife International.

Heimlich, R. E., K. D. Wiebe, R. Claassen, D. Gadsby, and R. M. House.1998. *Wetlands and Agriculture: Private Interests and Public Benefits.* Agriculture Economic Report No. 765. Washigton, DC: U.S. Department of Agriculture, Economic Research Service.

Hill, G., J. Waage, and G. Phiri. 1997. "The Water Hyacinth Problem in Tropical Africa," in E.S. Delfosse and N.R. Spencer, eds., *Proceedings of the International Water Hyacinth Consortium.* Washington, DC: World Bank. Available on-line at: http://www.sidney.ars.usda.gov/scientists/ nspencer/water_h/appendix5.htm

Hinrichsen, D., B. Robey, and U.D. Upadhyay. 1998. *Solutions for a Water-Short World.* Population Reports, Series M, No. 14. Baltimore, Maryland, U.S.A.: Johns Hopkins University School of Public Health, Population Information Program.

Houlahan, J. E., C. S. Findlay, B. R. Schmidt, A. H. Meyer, and S. L. Kuzmin. 2000. "Quantitative Evidence for Global Amphibian Population Declines." *Nature* 404 (6779): 752–755.

Hughes, R. M. and R. F. Noss. 1992. "Biological Diversity and Biological Integrity: Current Concerns for Lakes and Streams." *Fisheries*, May-June 1992, as cited in J. N. Abramovitz, *Imperiled Waters, Impoverished Future: The Decline of Freshwater Ecosystems.* Worldwatch Paper 128. Washington, DC: Worldwatch Insititute.

ICOLD (International Commission on Large Dams). 1998. *World Register of Dams 1998.* Paris, France: ICOLD.

IGBP (The International Geosphere-Biosphere Programme). 1998. *Global Wetland Distribution and Functional Characterization: Trace Gases and the Hydrologic Cycle.* Report from the Joint GAIM, BAHC, IGBP-DIS, IGAC, and LUCC Workshop, Santa Barbara, CA, U.S.A., 16–20 May 1996. D. Sahagian and J. Meybeck, eds. IGBP Report No. 46. Stockholm, Sweden: IGBP.

IJHD (International Journal of Hydropower and Dams). 1998. *1998 World Atlas and Industry Guide.* Surrey, U. K.: Aqua-Media International.

ISLCP (International Satellite Land Surface Climatology Project). 1987. "Global Data Set for Land-Atmosphere Models Initiative 1:87–88 Volumes 1–5." NASA Goddard Distributed Active Archive Center Science Data Series. Maryland, U.S.A.: NASA/Goddard Space Flight Center.

IUCN (The World Conservation Union). 1996. *1996 IUCN Red List of Threatened Animals.* Gland, Switzerland: IUCN-The World Conservation Union.

Jones, K. B., K. H. Riitters, J. D. Wickham, R. D. Tankersley Jr., R. V. O'Neill, D. J. Chaloud, E. R. Smith, and A. C. Neale. 1997. *An Ecological Assessment of the United States Mid-Atlantic Region: A Landscape Atlas.* Washington, DC: U.S. Environmental Protection Agency.

Karr, J. R. and E. W. Chu. 1999. *Restoring Life in Running Waters: Better Biological Monitoring.* Washington, DC: Island Press

Kapetsky, J. Inland Water Resources and Aquaculture Service, FAO Fisheries Department. 1999. Personal Communication. 27 August.

Kaufman, L. Boston University Marine Program. 2000. Personal Communication. Interview. 7 February.

Kaufman, L. 1992. "Catastrophic Change in Species-Rich Freshwater Ecosystems: The Lessons from Lake Victoria." *Bioscience* 42 (11): 846–858.

Kottelat, M. and T. Whitten. 1996. *Freshwater Biodiversity in Asia with Special Reference to Fish.* World Bank Technical Paper No. 343. Washington, DC: The World Bank.

Lelek, A., 1989. "The Rhine River and Some of its Tributaries Under Human Impact in the Last Two Centuries," pp. 469–487 in *Proceedings of the International Large River Symposium.* D. P. Dodge, ed. Canadian Special Publication of Fisheries and Aquatic Science 106. Ottawa, Canada: Department of Fisheries and Oceans.

Lerner, S. and W. Poole. 1999. *The Economic Benefits of Parks and Open Space: How Land Conservation Helps Communities Grow Smart and Protect the Bottom Line.* San Francisco, California, U.S.A.: The Trust for Public Land.

Liao, G. Z., K. X. Lu, and X. Z. Xiao. 1989. "Fisheries Resources of the Pearl River and Their Exploitation," pp. 561–568 in *Proceedings of the International Large River Symposium.* D. P. Dodge, ed. Canadian Special Publication of Fisheries and Aquatic Science 106. Ottawa, Canada: Department of Fisheries and Oceans.

Ligon, F. K., W. E. Dietrich, and W. J. Trush. 1995. "Downstream Ecological Effects of Dams: A Geomorphic Perspective." *Bioscience* 45 (3): 183–192.

Lips, K. R. 1998. "Decline of a Tropical Montane Amphibian Fauna." *Conservation Biology* 12 (1): 106–17.

L'vovich, M. I. and G. F. White. 1990. "Use and Transformation of Terrestrial Water Systems," pp. 235–252 in *The Earth as Transformed by Human Actions: Global and Regional Changes in the Biosphere Over the Past 300 Years.* B. L. Turner II, W. C. Clark, R. W. Kates, J. F. Richards, J. T. Mathews, and W. B. Meyer, eds. Cambridge, U. K.: Cambridge University Press.

Lyons, J., S. Navarro-Pérez, P. A. Cochran, E. Santana, and M. Guzmán-Arroyo. 1995. "Index of Biological Integrity Based on Fish Assemblages for the Conservation of Streams and Rivers in West-Central Mexico." *Conservation Biology* 9 (3): 569–584.

Mandal, B. K., T. R. Chowdhury, G. Samanta, G. K. Basu, P. K. Chowdhury, C. R. Chanda, D. Lodh, N. K. Karan, R. K. Dhar, D. K. Tamili, D. Das, K. C. Saha and D. Chakraborti. 1996. "Arsenic in Groundwater in Seven Districts of West Bengal, India—The Biggest Arsenic Calamity in the World." *Current Science* 70 (11): 976–86.

Marchant, R., A. Hirst, R. H. Norris, R. Butcher, L. Metzeling, and D. Tiller. 1997. "Classification and Prediction of Macroinvertebrate Assemblages from Running Waters in Victoria, Australia." *J. North American Bentholological Society* 16 (3): 664–81.

Marshbird Monitoring Website. Internet reference: http://www.mp1-pwrc.usgs.gov/marshbird/. Viewed 8/19/99.

Master, L. L., S. R. Flack, and B. A. Stein, eds. 1998. *Rivers of Life: Critical Watersheds for Protecting Freshwater Biodiversity.* Arlington, Virginia, U.S.A.: The Nature Conservancy.

Matthews, E. and I. Fung. 1987. "Methane Emission from Natural Wetlands: Global Distribution, Area, and Environmental Characteristics of Sources." *Global Biogeochemical Cycles* 1:61–86.

McAllister, D. E., A. L. Hamilton, and B. Harvey. 1997. "Global Freshwater Biodiversity: Striving for the integrity of freshwater ecosystems." *Sea Wind—Bulleting of Ocean Voice International* 11 (3): 1–140.

McCully, P. 1996. *Silenced Rivers: The Ecology and Politics of Large Dams.* London, U. K. and New Jersey, U.S.A.: Zed Books Ltd. and International Rivers Network.

Miller, R. R., J. D. Williams, and J. E. Williams. 1989. "Extinctions of North American Fishes During the Past Century." *Fisheries* 14 (6): 22–38.

Missouri River Coalition. 1995. "Comments on the Missouri River Water Control Manual Review and Update Draft Environmental Impact Assessment, March 1, 1995." As cited in J. N. Abramovitz, *Imperiled Waters, Impoverished Future: The Decline of Freshwater Ecosystems.* Worldwatch Paper 128. Washington, DC: Worldwatch Insititute.

Moyle, P.B. and P.J. Randall. 1998. "Evaluating the Biotic Integrity of Watersheds in the Sierra Nevada, California." *Conservation Biology* 12 (6): 1318–1326.

Moyle, P.B. and R.A. Leidy. 1992. "Loss of Biodiversity in Aquatic Ecosystems: Evidence from Fish Faunas," pp. 127–169 in *Conservation Biology: The Theory and Practice of Nature Conservation, Preservation and Management.* P.L. Fiedler and S.K. Jain, eds. New York, New York: Chapman and Hall.

MRC (Mekong River Commission). 1997. *Greater Mekong Sub-Region State of the Environment Report.* Bangkok, Thailand: Mekong River Commission and United Nations Environment Programme.

Mueller, D.K. and D.R. Helsel. 1996. *Nutrients in the Nation's Waters—Too Much of a Good Thing?* USGS Circular 1136. Reston, Virginia, U.S.A.: U.S. Geological Survey. Available on-line at: http://water.usgs.gov/nawqa/circ-1136/h10.html.

Myers, N. 1997. "The Rich Diversity of Biodiversity Issues," pp. 125–138 in *Biodiversity II: Understanding and Protecting Our Biological Resources,* M. L. Reaka-Kudla, D. E. Wilson, and E. O. Wilson, eds. Washington, DC: Joseph Henry Press.

Naiman, R. J., J. J. Magnuson, D. M. McKnight, and J. A. Stanford, eds. 1995. *The Freshwater Imperative: A Research Agenda.* Washington, DC: Island Press.

Naylor, R. L., R. J. Goldburg, J. H. Primavera, N. Kautsky, M. C. M. Beveridge, J. Clay, C. Folke, J. Lubchenco, H. Mooney, and M. Troell. 2000. "Effect of Aquaculture on World Fish Supplies." *Nature* 405 (6790): 1017–1024.

Nelson, J. S. 1976. *Fishes of the World.* New York, New York: Wiley.

Nelson, J. S. 1984 (2d ed.). *Fishes of the World.* New York, New York: Wiley.

Nelson, J. S. 1994 (3d ed.). *Fishes of the World.* New York, New York: Wiley.

NFNA (National Forest and Nature Agency). August 1997. *The Skjern River Restoration Project Interim Report.* Copenhagen, Denmark: Danish Ministry of the Environment and Energy.

NFNA (National Forest and Nature Agency). 1999. *The Skjern River Restoration Project booklet.* Copenhagen, Denmark: Danish Ministry of the Environment and Energy.

Ngantou, D. and R. Braund. 1999. "Waza-Logobe: Restoring the Good Life." *World Conservation* 30 (2): 19–20.

Nickson, R., J. MacArthur, W. Burgess, K.M. Ahmed, P. Ravenscroft, M. Rahman. 1998. "Arsenic Poisoning of Bangladesh Groundwater." *Nature* 395 (6700): 338.

Nilsson, C., M. Svedmark, P. Hansson, S. Xiong and K. Berggren. 2000. River fragmentation and flow regulation analysis. Unpublished data. Umeå, Sweden: Landscape Ecology, Umeå University.

NOAA-NGDC (National Oceanic and Atmospheric Administration-National Geophysical Data Center). 1998. *Stable Lights and Radiance Calibrated Lights of the World CD-ROM.* View Nighttime Lights of the World database available on-line at: http://julius.ngdc.noaa.gov:8080/production/html/BIOMASS/night.html. Boulder, Colorado, U.S.A.: NOAA-NGDC.

Nolan, B.T., B.C. Ruddy, K.J. Hitt, and D.R. Helsel. 1998. "A National Look at Nitrate Contamination of Ground Water." *Water Conditioning and Purification* 39 (12): 76–79.

NRC (National Research Council). 1992. *Restoration of Aquatic Ecosystems.* Washington, DC: National Academy Press.

Oberdorff, T., and R.M. Hughes. 1992. "Modification of an Index of Biotic Integrity Based on Fish Assemblages to Characterize Rivers of the Seine Basin, France." *Hydrobiologia* 228:11–130.

Oberdorff, T., J. F. Guégan, and B. Hugueny. 1995. "Global Scale Patterns of Fish Species Richness in Rivers." *Ecography,* 18: 345–352.

Oberdorff, T. 1997. Muséum National d'Histoire Naturelle, Lab. D'Ichtyologie Général et Appliquée, Paris, France. Unpublished data provided to WRI.

Olesen, K. W. and K. Havnr. 1998. *Restoration of the Skjern River: Towards a Sustainable River Management Solution.* Hørsholm, Denmark: Danish Hydraulic Institute.

Olson, D. and E. Dinerstein. 1999. *The Global 200: A Representation Approach to Conserving The Earth's Distinctive Ecoregions.* Draft Manuscript, March. Washington, DC: World Wildlife Fund (WWF-US).

O'Neill, C. R. 1996. *Economic Impact of Zebra Mussels: The 1995 National Zebra Mussel Information Clearinghouse Study.* Brockport, New York: New York Sea Grant Extension.

O'Neill, C. R. 1999. National Zebra Mussel Information Clearinghouse, New York Sea Grant. Personal Communication.

Pelley, Janet. 1998. "No Simple Answer to Recent Amphibian Declines." *Environmental Science and Technology* 32 (15): 352-53.

Postel, S. 1993. "Water and Agriculture," pp. 56–66 in *Water in Crisis: A Guide to the World's Fresh Water Resource,* P. Gleick, ed. New York, New York and Oxford, U.K.: Oxford University Press.

Postel, S. 1995. "Where Have All the Rivers Gone?" *World Watch* 8 (3): 9–19.

Postel, S. 1997. *Dividing the Waters.* Available on-line at: http://precision.uos.ac.kr/ leo/text/postel.html

Postel, S. 1999. *Pillar of Sand: Can the Irrigation Miracle Last?* Washington, DC: Worldwatch Institute.

Postel, S. and S. Carpenter. 1997. "Freshwater Ecosystem Services," pp. 195–214 in *Nature's Services: Societal Dependence on Natural Ecosystems.* G. C. Daily, ed. Washington, DC: Island Press.

Pounds, J. A., M. P. Fogden, J. M. Savage, and G. C. Gorman. 1997. "Tests of Null Models for Amphibian Declines on a Tropical Mountain." *Conservation Biology* 11: 1307–1322.

Ramsar Convention Website. 2000. Available on-line at: http://ramsar.org/key_sitelist.htm and Montreux Record at: http://ramsar.org/key_montreux_record.htm. Viewed 7/14/00.

Reaka-Kudla, M. L.1997. "The Global Biodiversity of Coral Reefs: A Comparison with Rain Forests," pp. 83–108 in *Biodiversity II: Understanding and Protecting Our Biological Resources,* M. L. Reaka-Kudla, D. E. Wilson, and E. O. Wilson, eds. Washington, DC: Joseph Henry Press.

Remane, K. ed. 1997. *African Inland Fisheries, Aquaculture and the Environment.* Food and Agriculture Organization of the United Nations (FAO). Rome, Italy: FAO.

Revenga, C., S. Murray, J. Abramovitz, and A. Hammond, 1998. *Watersheds of the World: Ecological Value and Vulnerability.* Washington, DC: World Resources Institute.

Riber, H. H. COWI Consulting Engineers and Planners AS. 2000. Personal Communication. 16 January.

Ricciardi, A. and J. B. Rasmussen. 1999. "Extinction Rates of North American Freshwater Fauna." *Conservation Biology* 13 (5): 1220–1222.

Ross, S. T. 1991. "Mechanisms Structuring Stream Fish Assemblages: Are There Lessons From Introduced Species?" *Environmental Biology of Fishes* 30: 359–368.

Sauer, J. R., J. E. Hines, G. Gough, I. Thomas, and B.G. Peterjohn. 1997. *The North American Breeding Bird Survey Results and Analysis.* Version 96.3. Laurel, Maryland, U.S.A: Patuxent Environmental Science Center.

Scheidleder, J. Grath, G. Winkler, U. St@rk, C. Koreimann, and C. Gmeiner. 1999. *Groundwater Quality and Quantity in Europe.* S. Nixon, ed. European Topic Centre on Inland Waters. Copenhagen, Denmark: European Environment Agency.

Schloesser, Don W., and Thomas F. Nalepa. 1994. "Dramatic Decline of Unionid Bivalves in Offshore Waters of Western Lake Erie after Infestation by the Zebra Mussel, *Dreissena polymorpha.*" *Canadian Journal of Fish and Aquatic Science*, 51: 2234–2242.

SEAC (State of the Environment Advisory Council). 1996. *Australia State of the Environment Report,* Australian Department of the Environment Sport and Territories. Collingwood, Australia: CSIRO Publishing.

Seckler, D., U. Amarasinghe, D. Molden, R. de Silva, and R. Barker. 1998. *World Water Demand and Supply, 1990 to 2025: Scenarios and Issues.* Research Report 19. Colombo, Sri Lanka: International Water Management Institute.

Seckler, D. 1999. International Water Management Institute, Colombo, Sri Lanka. Personal Communication. 15 May.

Shiklomanov, I.A. 1997. *Comprehensive Assessment of the Freshwater Resources of the World: Assessment of Water Resources and Water Availability in the World.* Stockholm, Sweden: World Meteorological Organization and Stockholm Environment Institute.

Smith, R.A., R.B. Alexander, and K.J. Lanfear. 1994. *Stream Water Quality in the Coterminous United States—Status and Trends of Selected Indicators During the 1980s.* USGS Water Supply Paper 2400. Reston, Virginia, U.S.A.: U.S. Geological Survey.

Sparks, R. E. 1992. "The Illinois River Floodplain Ecosystem," in National Research Council, *Restoration of Aquatic Ecosystems: Science, Technology and Public Policy,* Washington, DC: National Academy Press. As cited in J. N. Abramovitz, *Imperiled Waters, Impoverished Future: The Decline of Freshwater Ecosystems.* Worldwatch Paper 128. Washington, DC: Worldwatch Insititute.

Taylor, R. and I. Smith. 1997. *State of New Zealand's Environment 1997.* Wellington, New Zealand: The Ministry for the Environment. Available on-line at: http://www.mfe.govt.nz/about/publications/ser/front.pdf.

TERI (Tata Energy Research Institute). 1998. *Looking Back to Think Ahead: Green India 2047.* New Delhi, India: TERI.

The Nature Conservancy, 1997. Natural Heritage Central Databases, The Nature Conservancy and the International Network of Natural Heritage Programs and Conservation Data Centers. November 1997. Available on-line at: http://www.consci.tnc.org/src/zoodata.htm.

Tucker, G. M., and M. I. Evans, 1997. *Habitats for Birds in Europe: A Conservation Strategy for the Wider Environment.* No.6, BirdLife Conservation Series, Duncan Brooks, ed. Cambridge, U.K.: BirdLife International.

Tyler, M. J. 1997. *The Action Plan for Australian Frogs.* Adelaide, Australia: Wildlife Australia Endangered Species Program, The University of Adelaide.

UN (United Nations). 1997. *China: Water Resources and Their Use.* Economic and Social Commission for Asia and the Pacific. New York, New York: United Nations.

UN/ECE (United Nations Economic Commission for Europe). 1995. *State of the Art on Monitoring and Assessment: Rivers.* UN/ECE Task Force on Monitoring and Assessment, Draft Report V. Lelystad, the Netherlands: RIZA.

UNEP (United Nations Environment Programme). 1996. *Groundwater: A Threatened Resource.* UNEP Environment Library No. 15. Nairobi, Kenya: UNEP.

UNEP/GEMS (United Nations Environment Program Global Environment Monitoring System/Water). 1995. *Water Quality of World River Basins.* Nairobi, Kenya: UNEP.

UNESCO (United Nations Educational, Scientific, and Cultural Organization). 2000. *Water Related Vision for the Aral Sea Basin for the Year 2025.* Paris, France: UNESCO.

UNPD (United Nations Population Division). 1999. *World Population Prospects: The 1998 Revision.* Vol. 1. New York, New York: United Nations.

USFWS (U.S. Fish and Wildlife Service). 1996. *National Survey of Fishing, Hunting, and Wildlife-Associated Recreation.* Washington, DC: U.S. Department of the Interior, Fish and Wildlife Service and U.S. Department of Commerce, Bureau of Census.

USGS (U.S. Geological Survey). Zebra Mussel Website. Available on-line at: http://nas.er. usgs.gov/zebra.mussel/docs/sp_account.html#HDR1. Viewed 7/19/00.

USGS (U.S. Geological Survey). Invasive Species Website. Available on-line at: http://nas.er.usgs.gov/plants/maps/smec.gif. Viewed 7/19/00.

USGS (U.S. Geological Survey). Landsat 7 Web Site. Available on-line at: http://landsat7.usgs.gov. Viewed 7/30/00.

USGS (U.S. Geological Survey). 1999. *The Quality of Our Nation's Waters — Nutrients and Pesticides.* USGS Circular 1225. Reston, Virginia, U.S.A.: USGS.

Vörösmarty, C. J., K. P. Sharma, B. M. Fekete, A. H. Copeland, J. Holden, J. Marble, and J. A. Lough. 1997a. "The Storage and Aging of Continental Runoff in Large Reservoir Systems of the World." *Ambio* 26(4): 210–219.

Vörösmarty, C. J., M. Meybeck, B. M. Fekete, and K. P. Sharma. 1997b. "The Potential Impact of Neo-Castorization on Sediment Transport by the Global Network of Rivers," pp. 261–272 in D. Walling and J. L. Probst, eds. *Human Impact on Erosion and Sedimentation.* Wallingford, U. K.: IAHS Press.

Vörösmarty, C., and D. Sahagian. 2000. "Anthropogenic Disturbance of the Terrestrial Water Cycle." *Bioscience*. In press.

Wager, R. and P. Jackson. 1993. *The Action Plan for Australian Freshwater Fishes.* Australian Nature Conservation Agency (ANCA), Endangered Species Program, Project No. 147. Canberra, Australia: ANCA.

Ward, J. V. and J. A. Stanford. 1989. "Riverine Ecosystems: The Influence of Man on Catchment Dynamics and Fish Ecology," pp. 56–64 in D. P. Dodge, ed. *Proceedings of the International Large River Symposium*, Canadian Special Publication of Fisheries and Aquatic Sciences 106. Ottawa, Canada: Department of Fisheries and Oceans.

Washington Department of Fish and Wildlife and Oregon Department of Fish and Wildlife. 1999. *Status Report Columbia River Fish Runs and Fisheries, 1938-1998.* Vancouver, Washington and Clackamas, Oregon, U.S.A.: Joint Columbia River Management Staff.

Watson, R. T., M. C. Zinyowera, and R. H. Moss. 1996. *Climate Change 1995, Impacts, Adaptations and Mitigation of Climate Change: Scientific Technical Analyses.* Contribution of Working Group II to the Second Assessment Report of the Intergovernmental Panel on Climate Change. Cambridge, U.K.: Cambridge University Press.

WCMC (World Conservation Monitoring Centre). 1996. *Biodiversity Map Library.* Cambridge, U.K.: WCMC.

Welcomme, R.L. 1988. *International Introductions of Inland Aquatic Species.* Food and Agriculture Organization of the United Nations (FAO) Technical Series Paper 294. Rome, Italy: FAO.

WHO (World Health Organization). 1996. Water and Sanitation Fact Sheet. Online at: http://www.who.org/inf-fs/en/fact112.html. Viewed September 12, 2000.

Williams, J. E. Johnson, D. A. Hendrickson, S. Contreras-Balderas, J. D. Williams, M. Navarro-Mendoza, D. E. McAllister, and D. E. Deacon. 1989. "Fishes of North America Endangered, Threatened or of Special Concern: 1989." *Fisheries* 14(6): 2–20.

Witte, F., T. Goldschmidt, J. Wanink, M. van Oijen, K. Goudswaard, E. Witte-Mass, and N. Bouton. 1992. "The Destruction of an Endemic Species Flock: Quantitative Data on the Decline of the Haplochromine Cichlids of Lake Victoria." *Environmental Biology of Fishes* 34:1-28.

WMO (World Meteorological Organization). 1997. *Comprehensive Assessment of the Freshwater Resources of the World.* Stockholm, Sweden: WMO and Stockholm Environment Institute.

Wood, S., K. Sebastian, and S. J. Scherr. 2000. *Pilot Analysis of Global Ecosystems: Agroecosystems Technical Report.* Washington, DC: World Resources Institute and International Food Policy Research Institute.

WRI (World Resources Institute). 1995. *Africa Data Sampler User's Guide; A Georeferenced Database for all African Countries.* Washington, DC: WRI.

WRI (World Resources Institute) in collaboration with the United Nations Environment Programme, The United Nations Development Programme, and the World Bank 1996. *World Resources 1996-97.* New York, New York: Oxford University Press.

Wright, J .F. 1995. "Development and Use of a System for Predicting the Macroinvertebrate Fauna in Flowing Waters." *Australian Journal of Ecology* 20: 181–97.

Yoder, C. O. 1991. "Answering Some Concerns about Biological Criteria Based on Experiences in Ohio," pp. 95–104 in *Proceedings: Water Quality Standards for the 21st Century*. Washington, DC: US Environmental Protection Agency, Criteria and Standards Division.

Yoder, C. O. and E. T. Rankin. 1995. "Biological Criteria Program Development and Implementation in Ohio," chapter 9 in W.S. Davis and T. Simon, eds. *Biological Assessment and Criteria: Tools for Water Resource Planning and Decision Making*. Boca Raton, Florida, U.S.A: Lewis Publishers.

Yoder, C. O., and E. T. Rankin. 1998. "The Role of Biological Indicators in a State Water Quality Management Process." *Environmental Management and Assessment* 51: 61–88.

Zhang, W.L., Z.X. Tian, N. Zhang, and X.Q. Li. 1996. "Nitrate Pollution of Groundwater in Northern China." *Agriculture, Ecosystems and Environment* 59: 223-31.

Map 1
River Channel Fragmentation and Flow Regulation

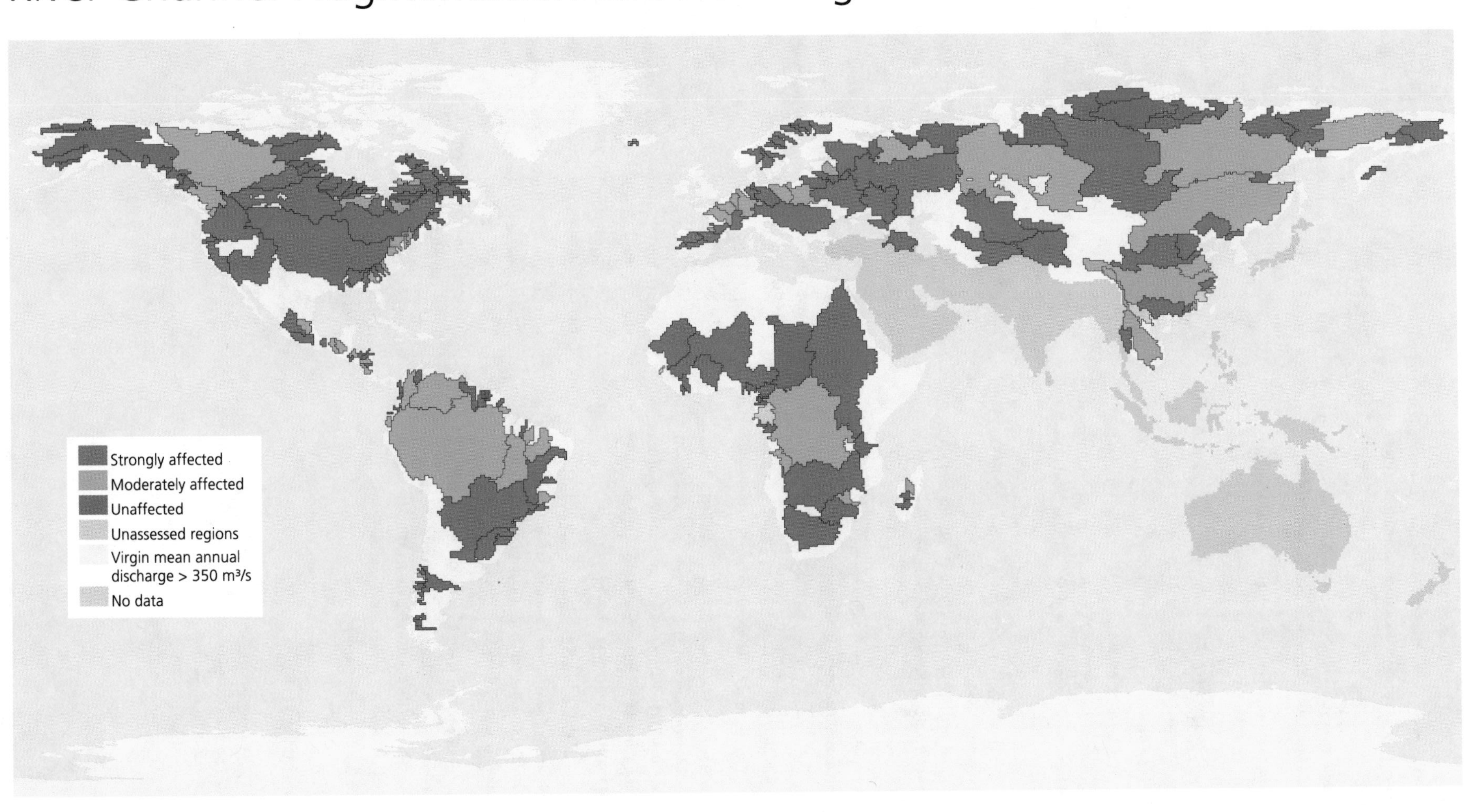

Source: Dynesius and Nilsson, 1994 and Nilsson et al., 2000. Watershed boundaries are from Fekete et al., 1999.
Projection: Geographic

Map 2
Large Dams under Construction by River Basin as of 1998

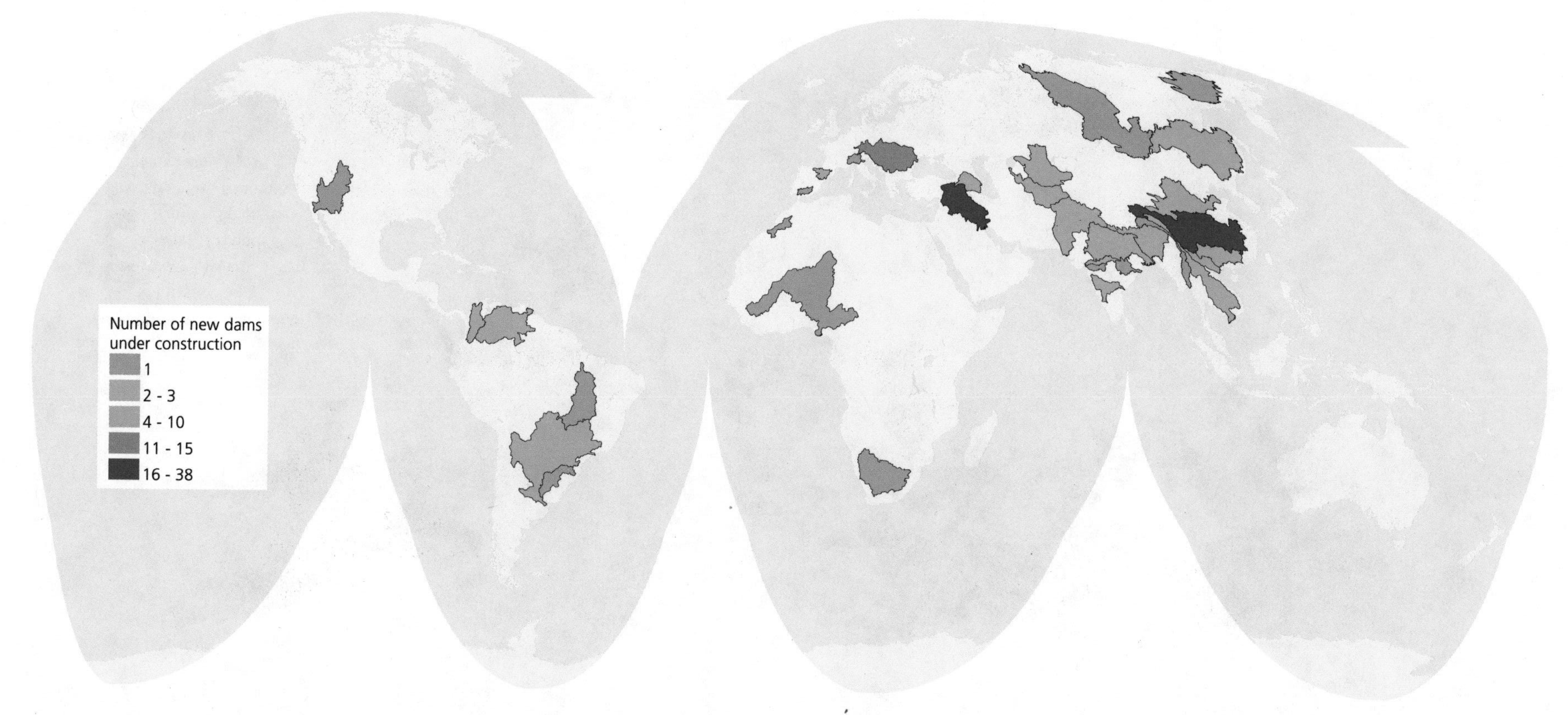

Source: IJHD, 1998. Watershed boundaries are from Revenga et al., 1998.
Projection: Interrupted Goode's Homolosine
Note: Only major river basins are shown on this map.

Map 3
Residence Time of Continental Runoff by River Basin

Source: Vörösmarty et al. 1997a. Watershed boundaries are from Fekete et al., 1999.
Projection: Geographic

Map 4
Africa: Wetlands, Dams, and Ramsar Sites

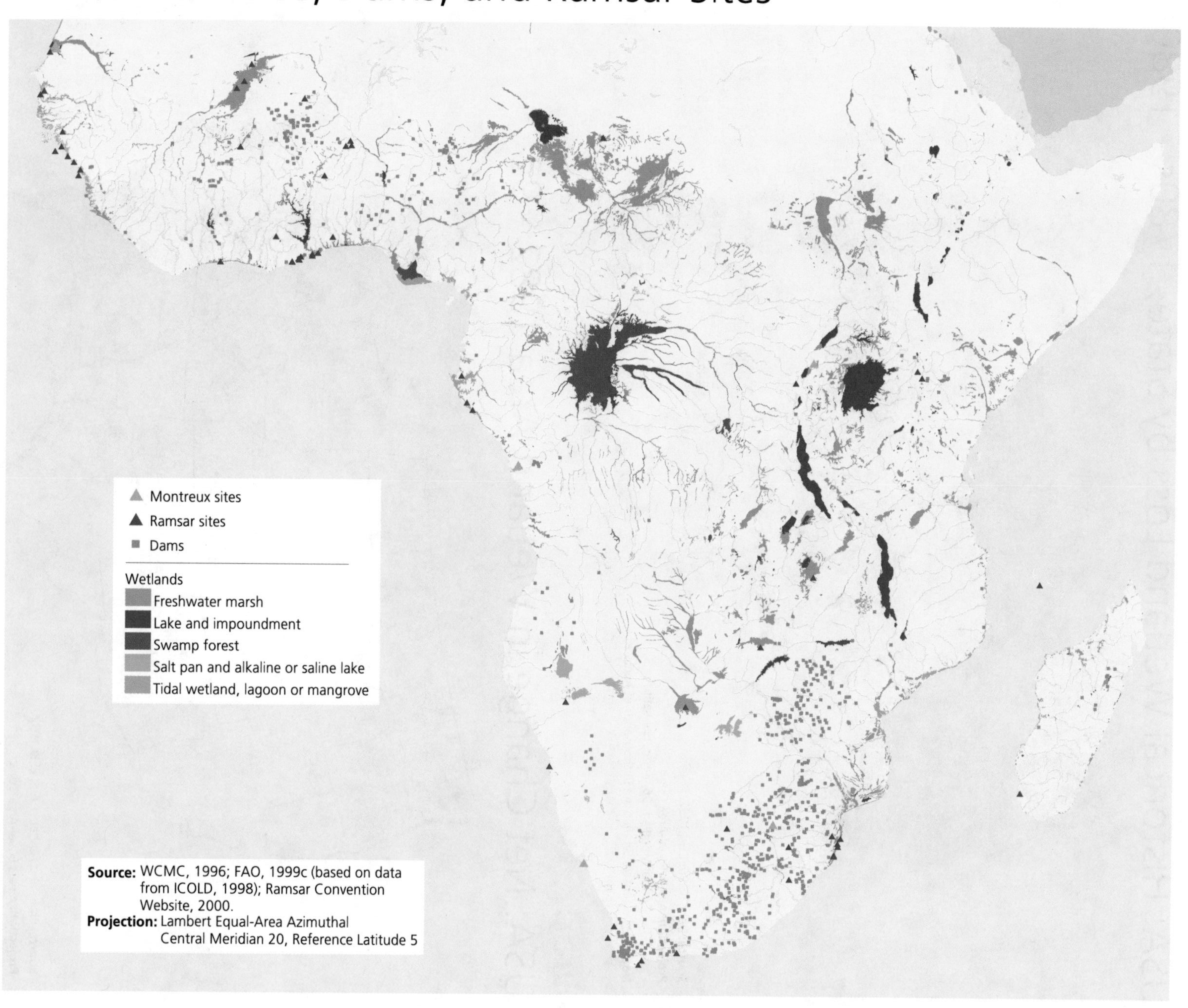

Source: WCMC, 1996; FAO, 1999c (based on data from ICOLD, 1998); Ramsar Convention Website, 2000.
Projection: Lambert Equal-Area Azimuthal Central Meridian 20, Reference Latitude 5

Map 5a

USA: Historical Wetland Loss by State, 1780s - 1980s

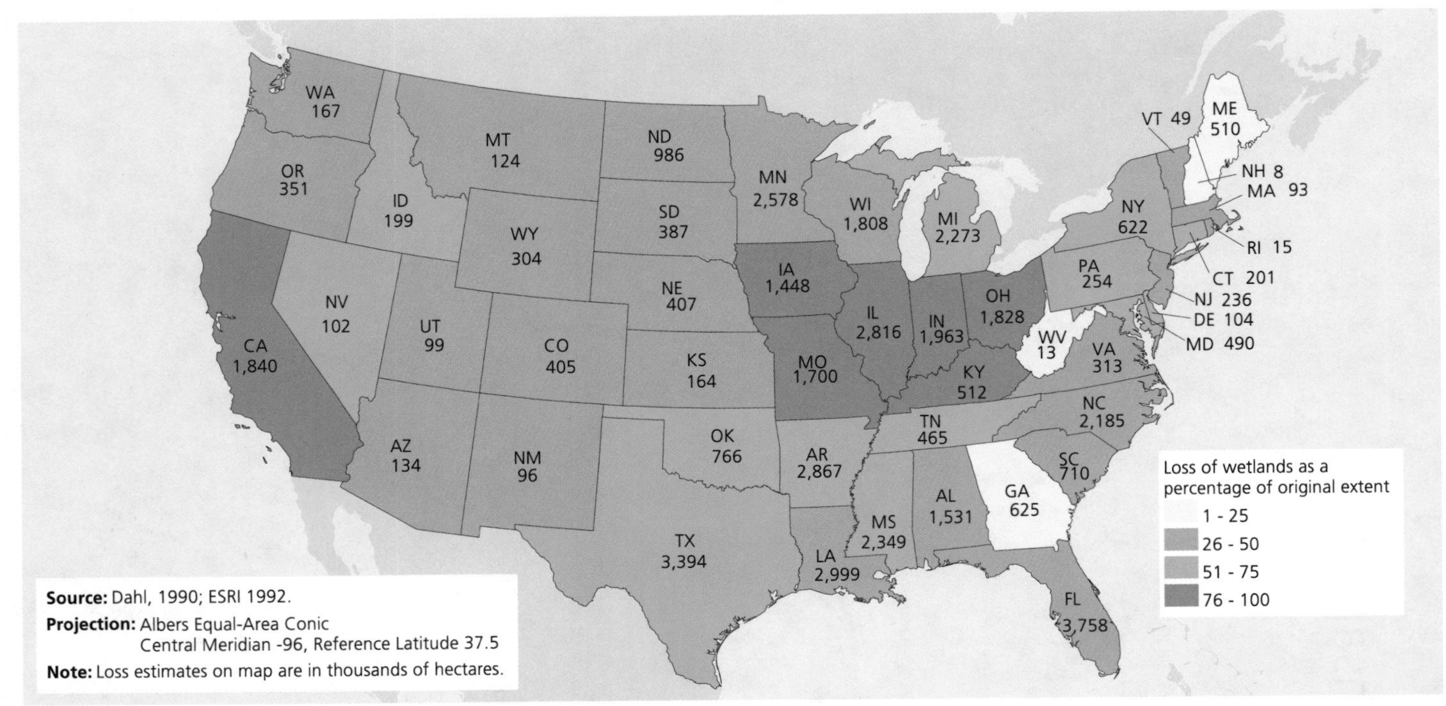

Map 5b

USA: Net Change in Wetland Area 1982 - 92

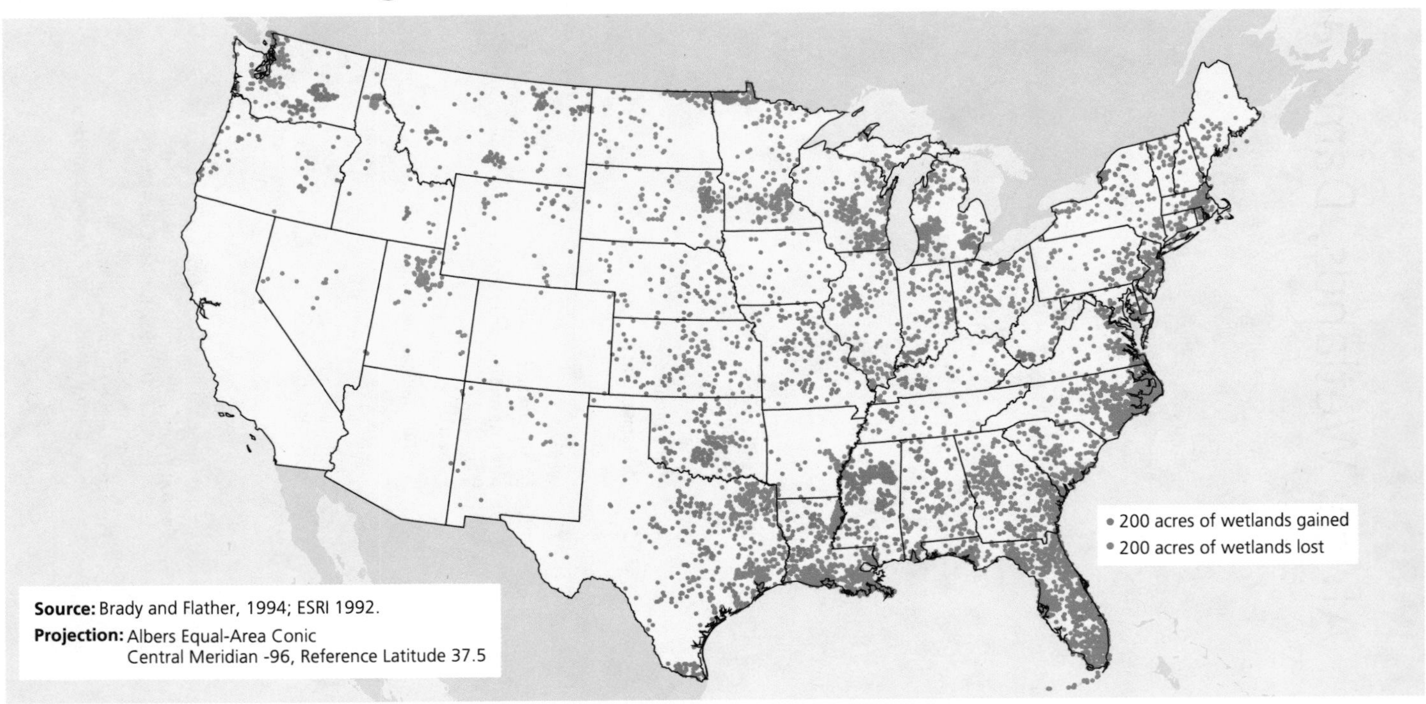

Map 6
Percentage of Cropland Area by River Basin

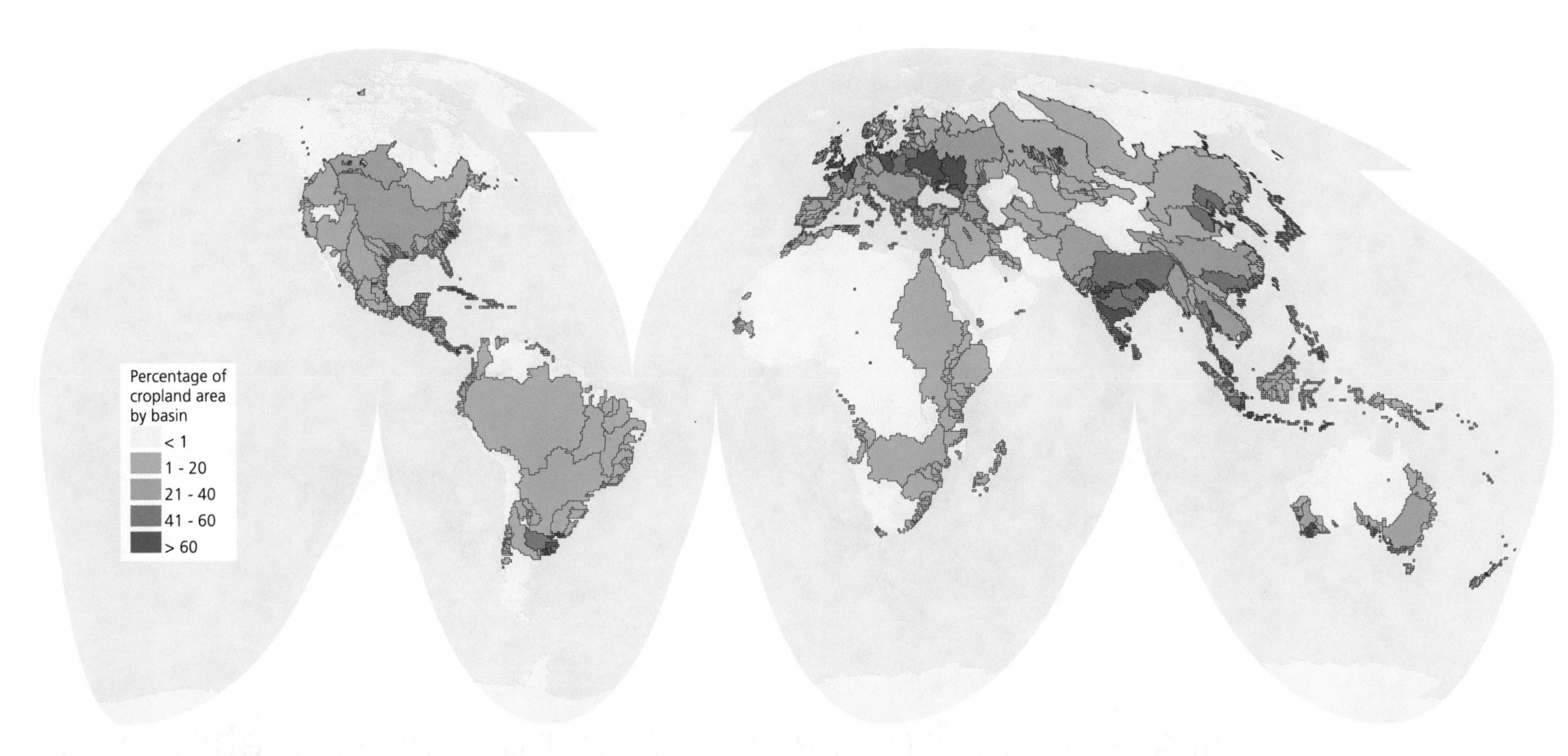

Source: GLCCD, 1998. Watershed boundaries are from Fekete et al., 1999.
Projection: Interrupted Goode's Homolosine

Map 7
Percentage of Urban and Industrial Land Use by River Basin

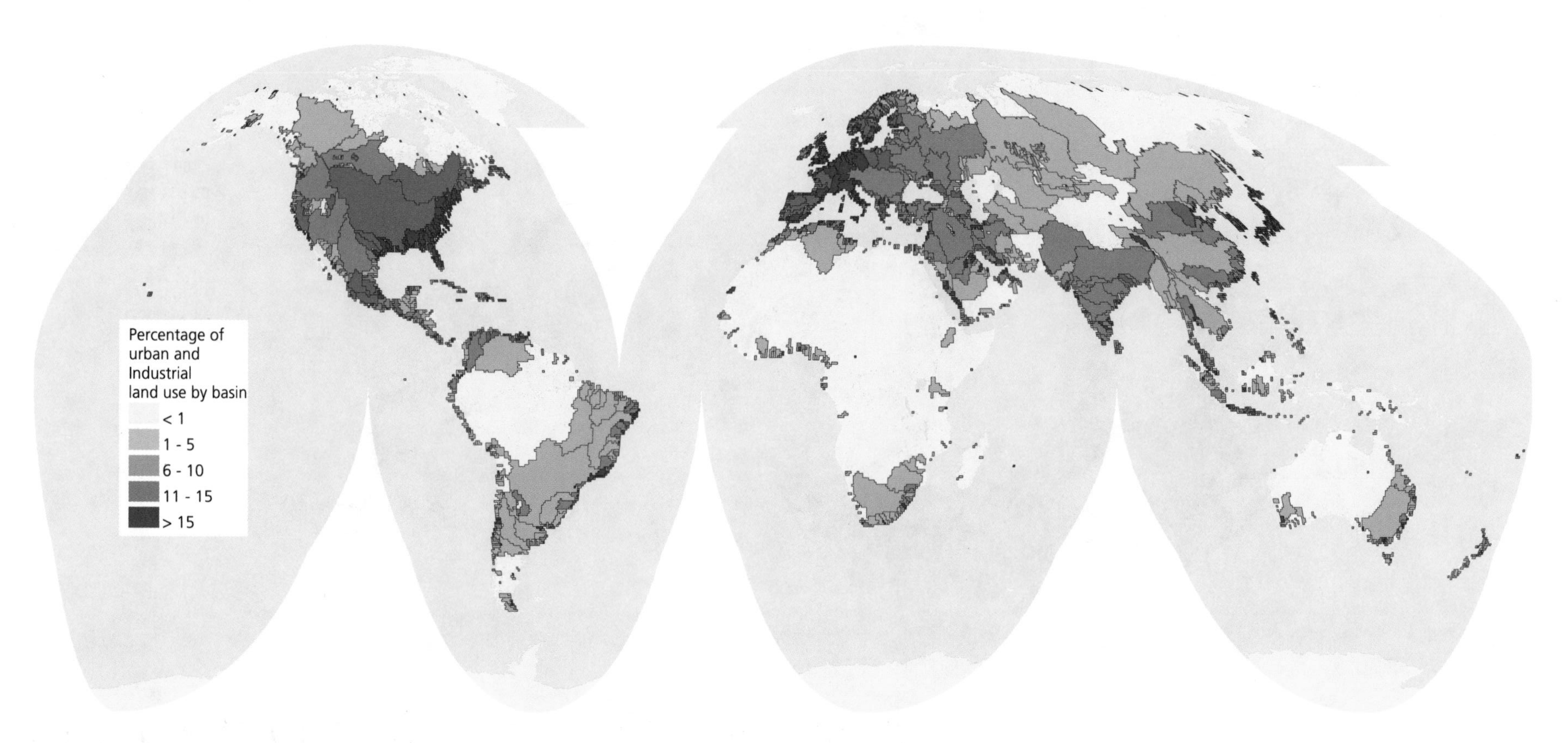

Source: NOAA/NGDC, 1998. Watershed boundaries are from Fekete et al., 1999.
Projection: Interrupted Goode's Homolosine

Maps 8a and 8b
Intensive Agricultural Land Use by Subbasin

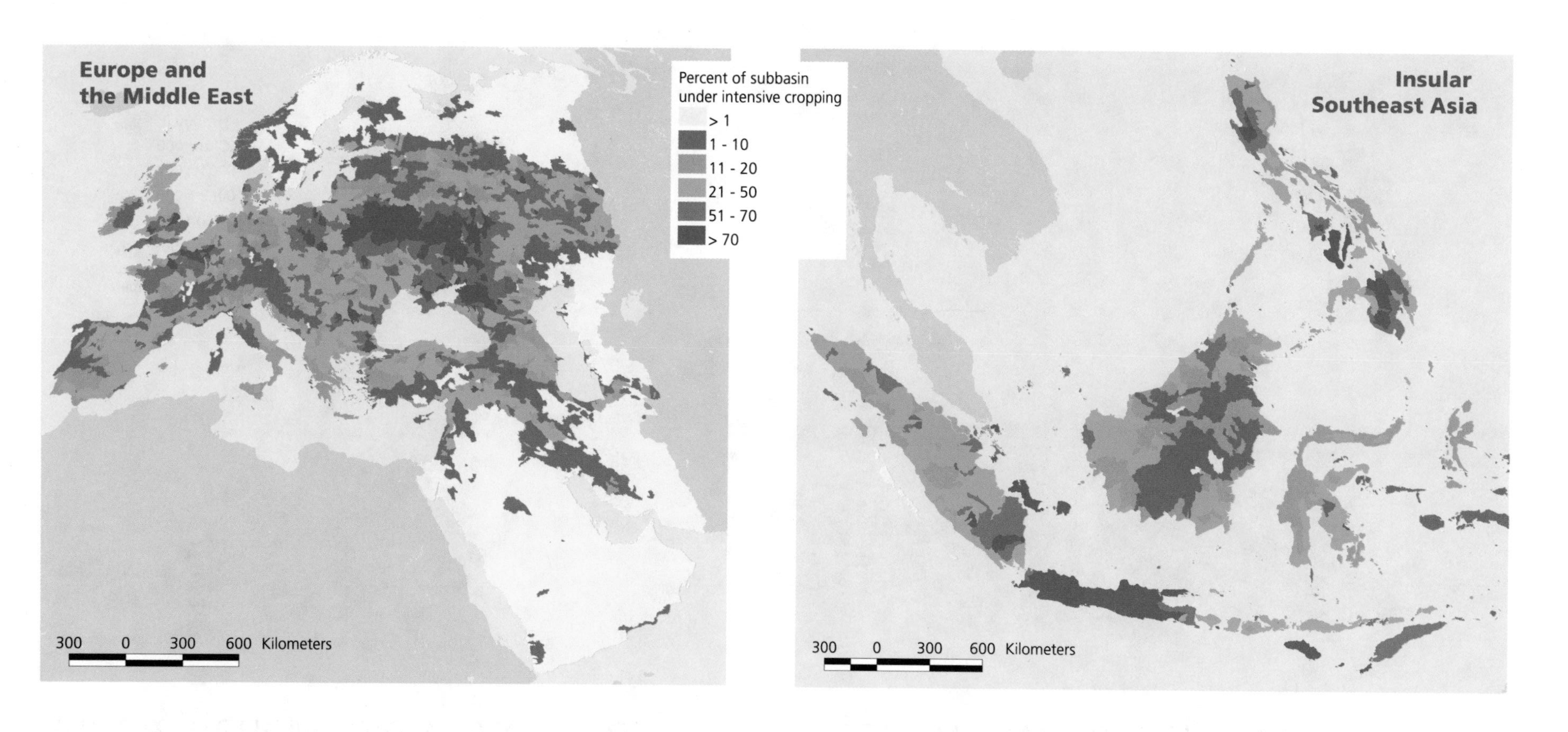

Source: GLCCD, 1998. Watershed boundaries are from EDC 1999.
Projection: Interrupted Goode's Homolosine

Map 9
Annual Renewable Water Supply Per Person by River Basin, 1995

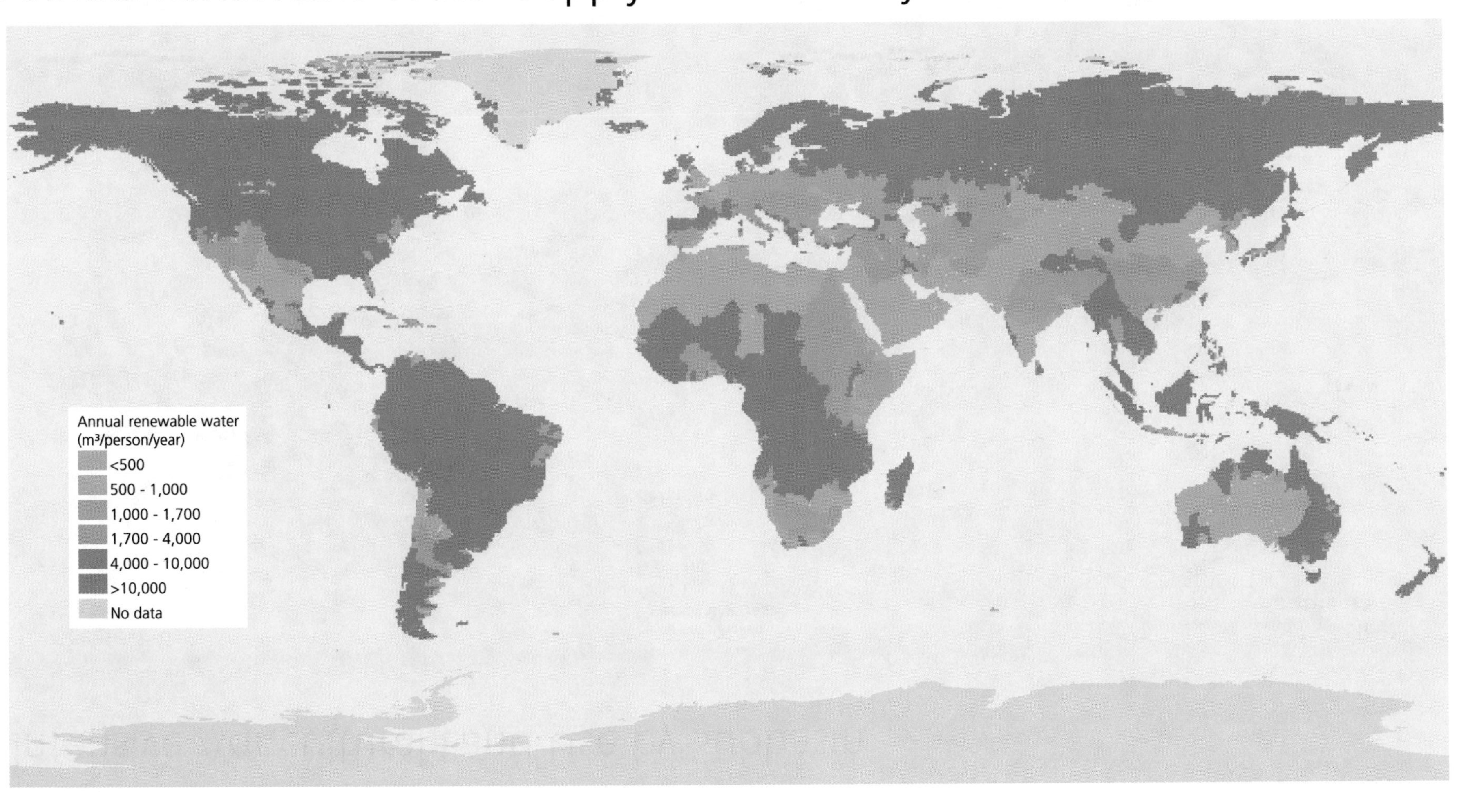

Source: CIESIN et al., 2000; Fekete et al., 1999.
Projection: Geographic

Map 10

Projected Annual Renewable Water Supply Per Person by River Basin, 2025

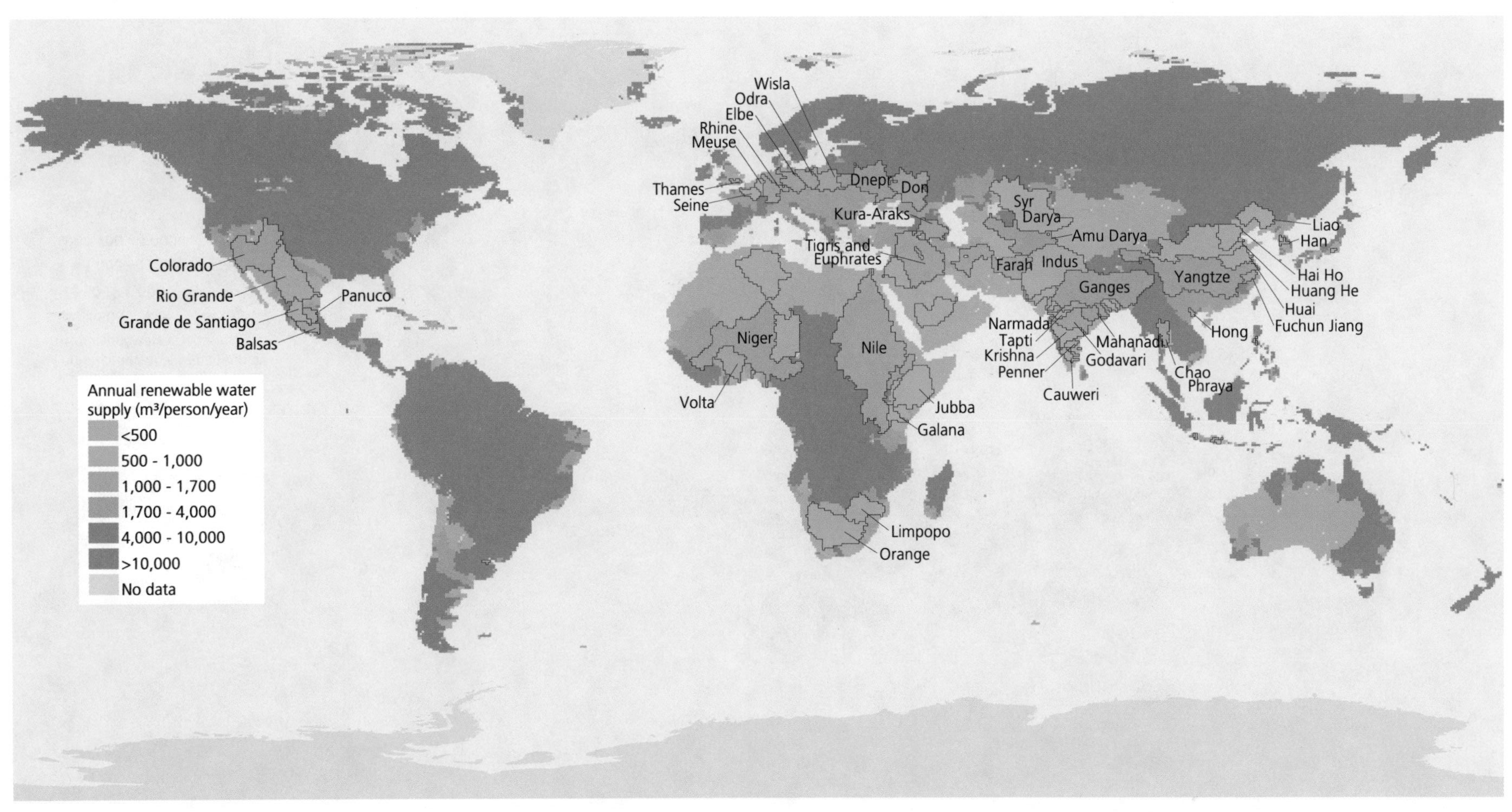

Source: CIESIN et al., 2000; Fekete et al., 1999.

Projection: Geographic

Note: Outlined basine are projected to have a population of more than 10,000,000 people in 2025. These basins are also in or approaching water scarcity, with less than 2,500 m^3 of water per person per year. Unlabeled, outlined basins in Africa and the Middle East have no perennial river flowing through them.

Map 11

Annual Renewable Water Supply and Dry Season Flow by River Basin

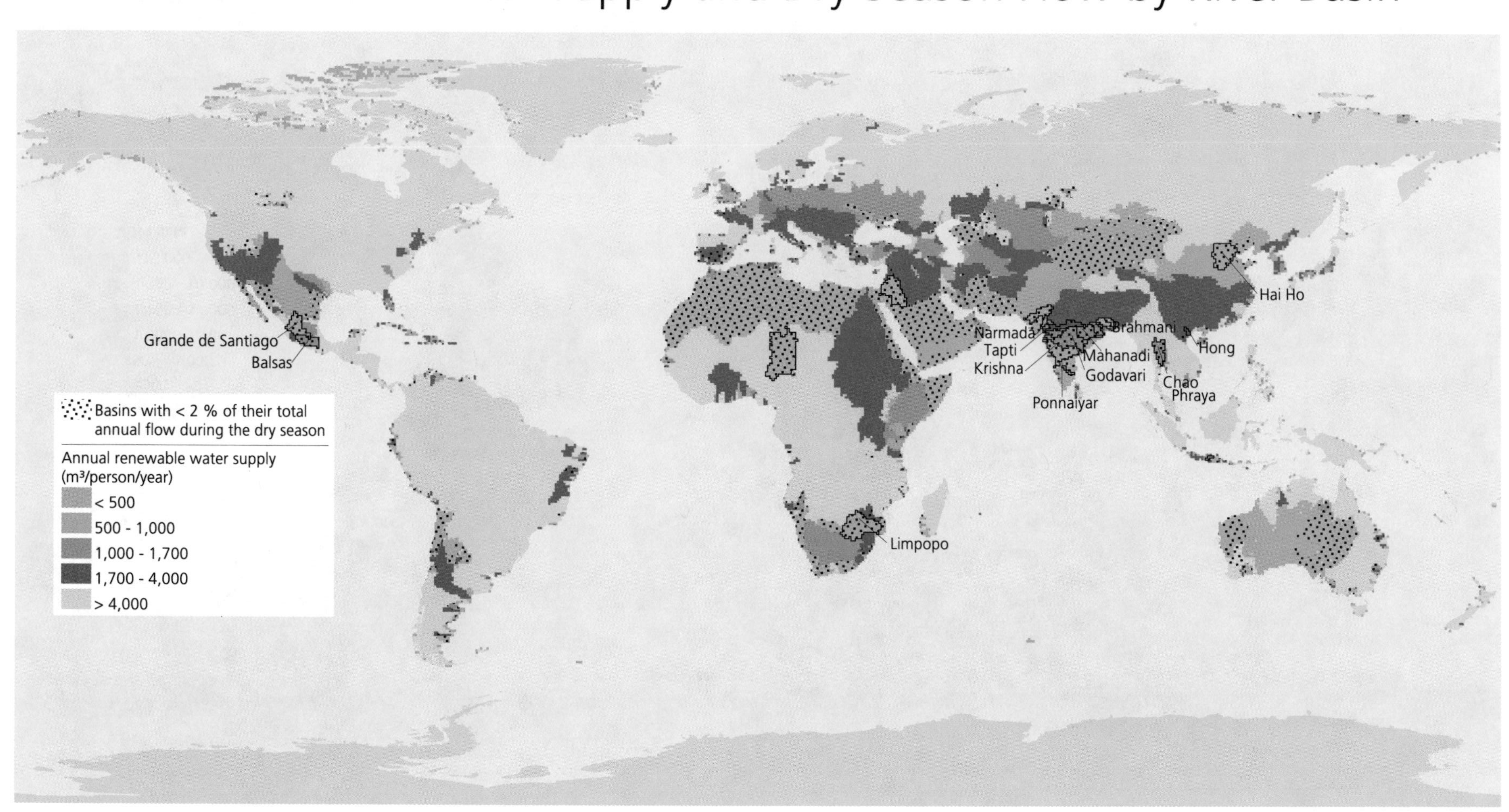

Source: Fekete et al., 1999; CIESIN et al., 2000.

Projection: Geographic

Note: Outlined, labeled basins are those which have both less than 2 percent of their total annual flow during the dry season and an estimated 1995 population of greater than 10,000,000 people. Outlined, unlabeled basins in Africa and the Middle East meet the above conditions, but have no perennial river flowing through them.

Map 12
Trends in Inland Capture Fisheries by Country, 1984 - 97

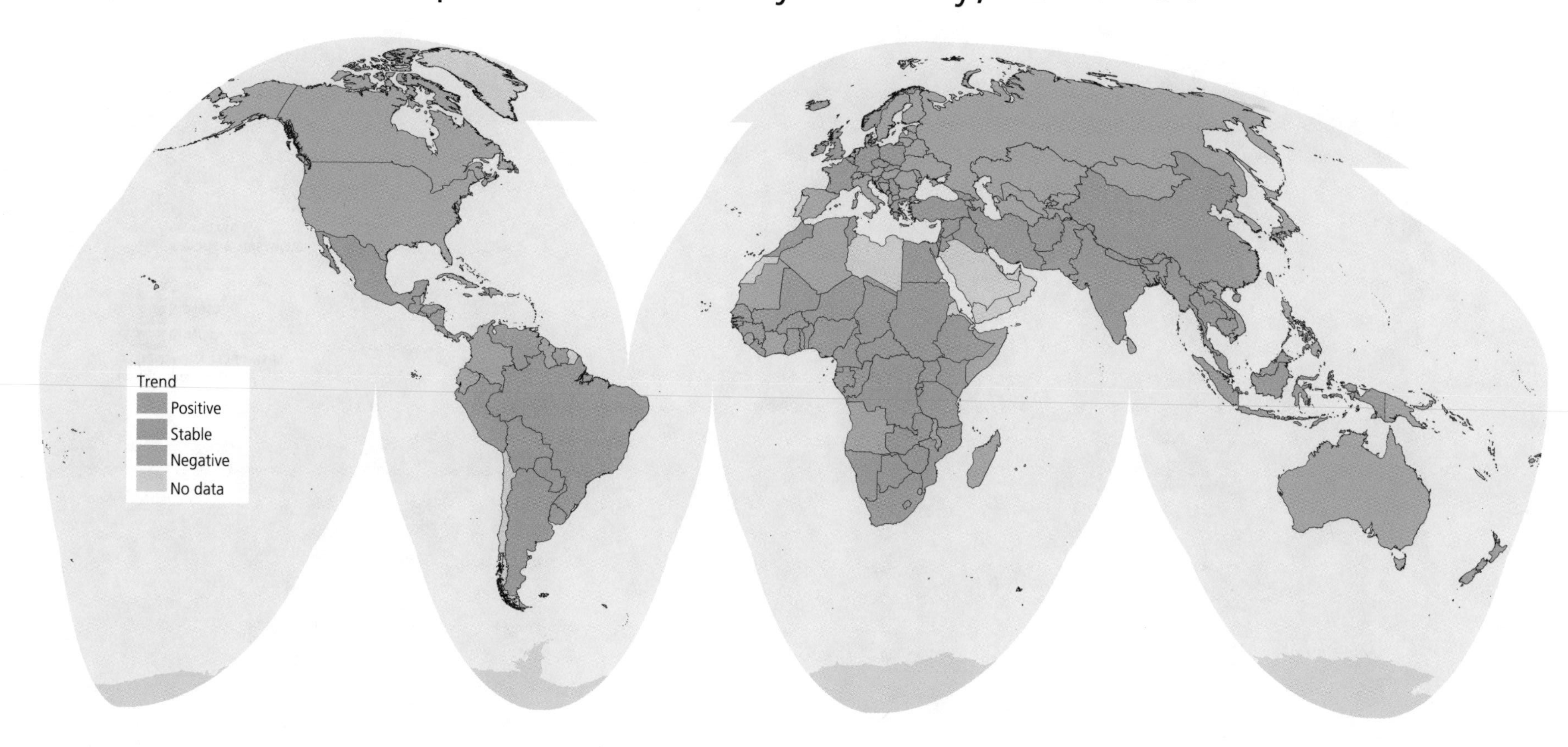

Source: FAO, 1999a; ESRI, 1996.
Projection: Interrupted Goode's Homolosine

Map 13
Important Areas and Ecoregions for Freshwater Biodiversity

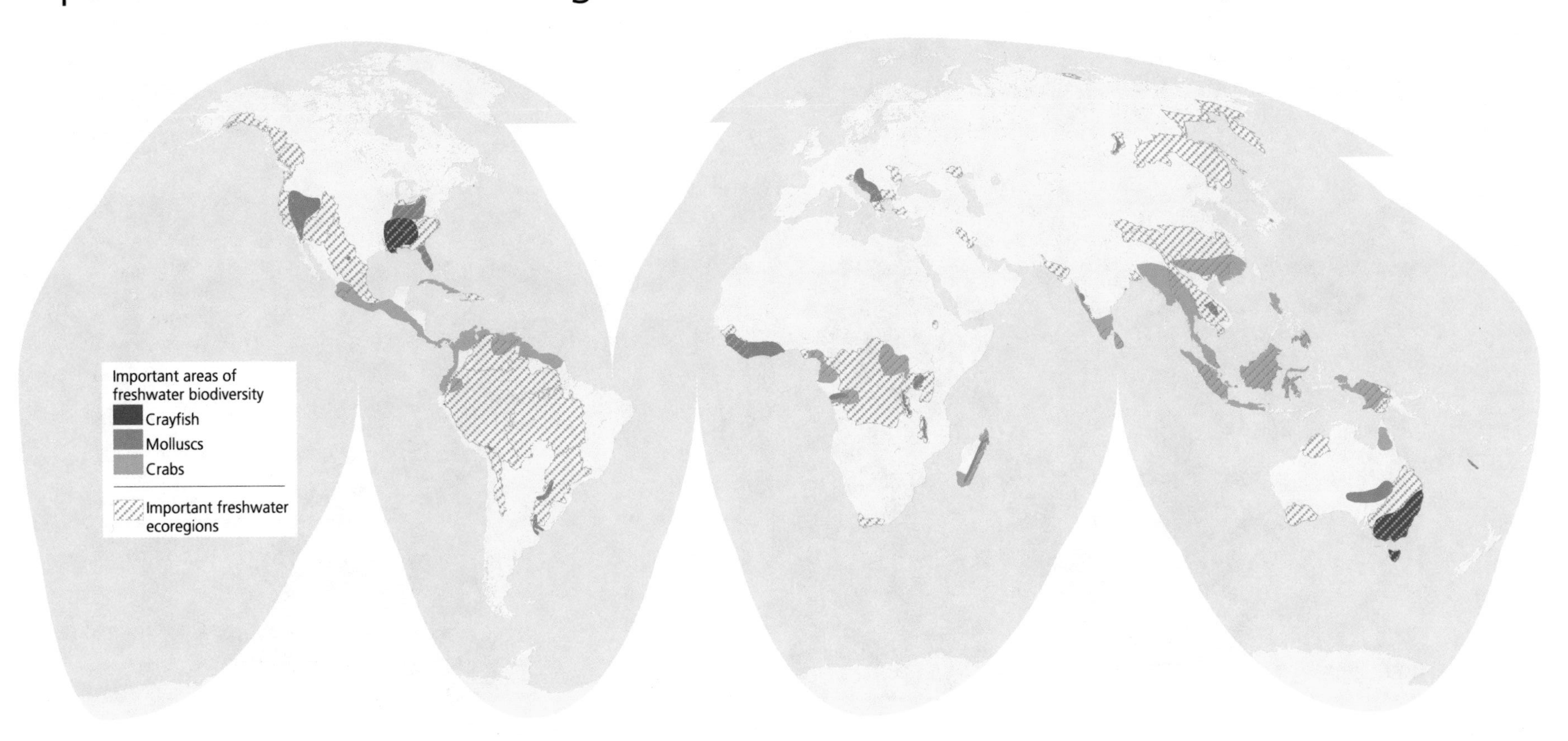

Source: Olson and Dinerstein, 1999; Goombridge and Jenkins, 1998.

Projection: Interrupted Goode's Homolosine

Note: Important areas for molluscs, crayfish, and crabs are from Goombridge and Jenkins (1998) based on data provided by P. Bouchet, O. Gargominy, A. Borgan, W. Ponder, K. Crandall, N. Cumberlidge, R. von Sternberg, D. Belk, and the IUCN/Species Survival Commission Inland Water Crustacean and Mollusc Specialist Groups.

Map 14
Fish Species Richness and Endemism by River Basin

Source: Revenga et al., 1998.
Projection: Interrupted Goode's Homolosine

Map 15
Amphibian Census Sites and Decline Index

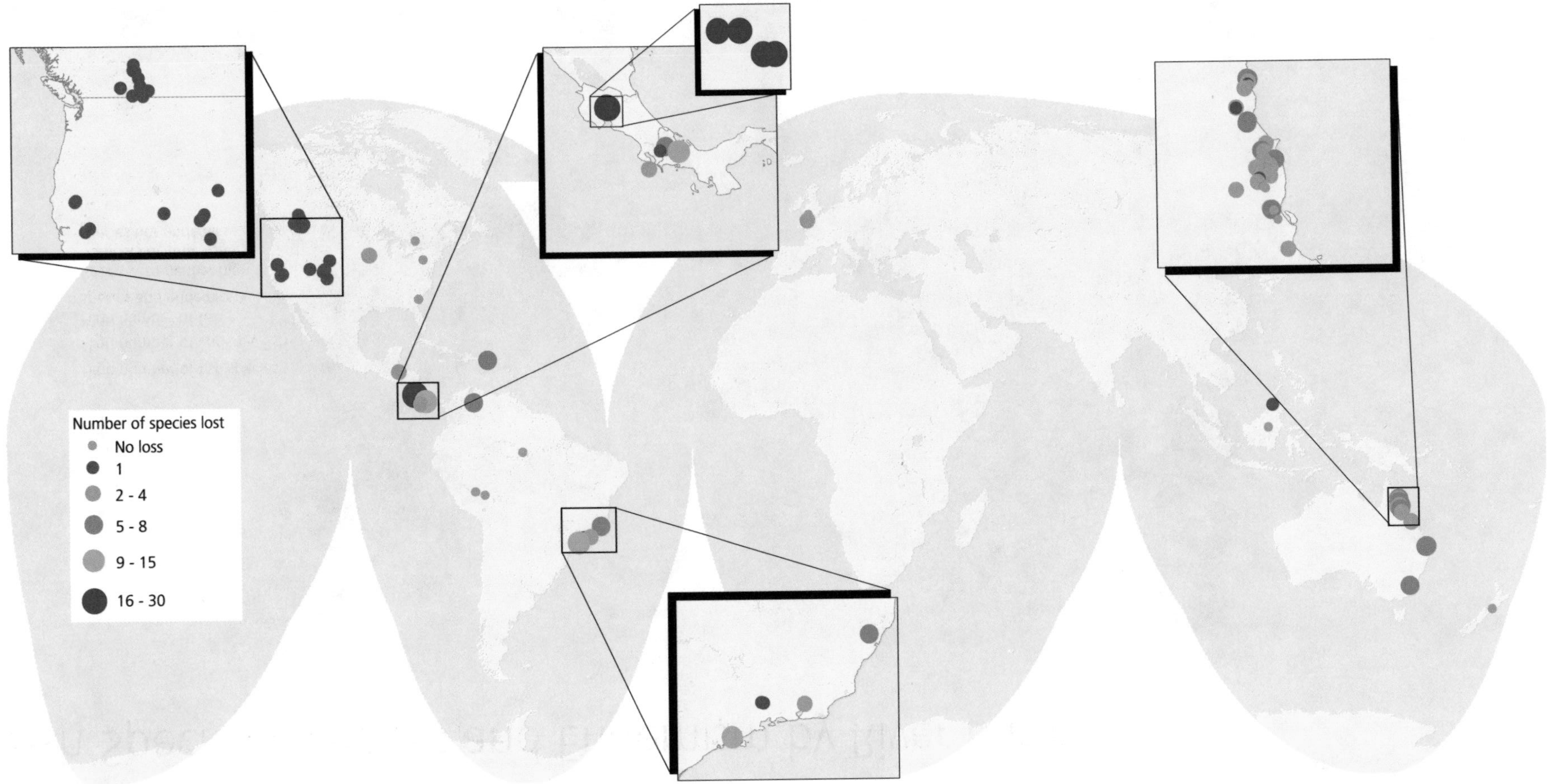

Sources: Carey et al., 2000
Projection: Interrupted Goode's Homolosine

Map 16
Imperiled Fish and Herpetofauna in North American Freshwater Ecoregions

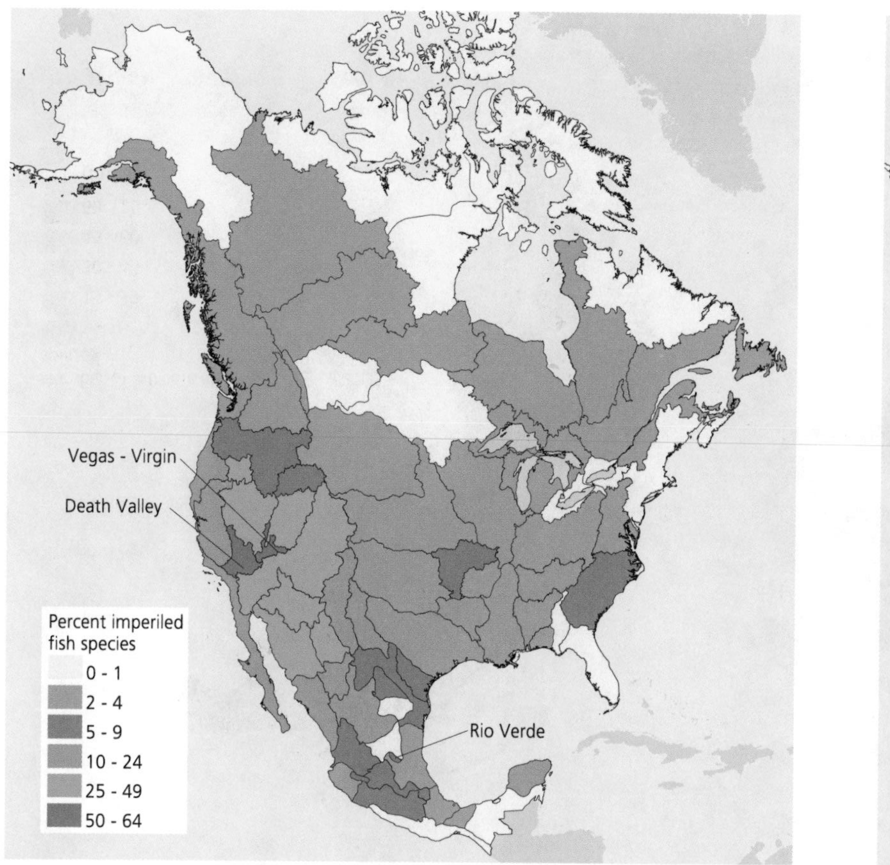

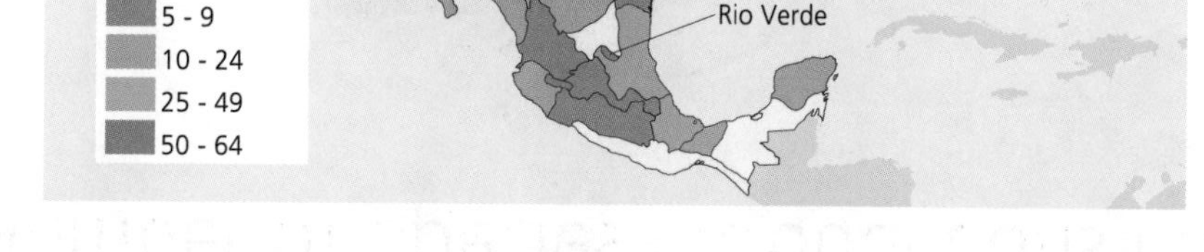

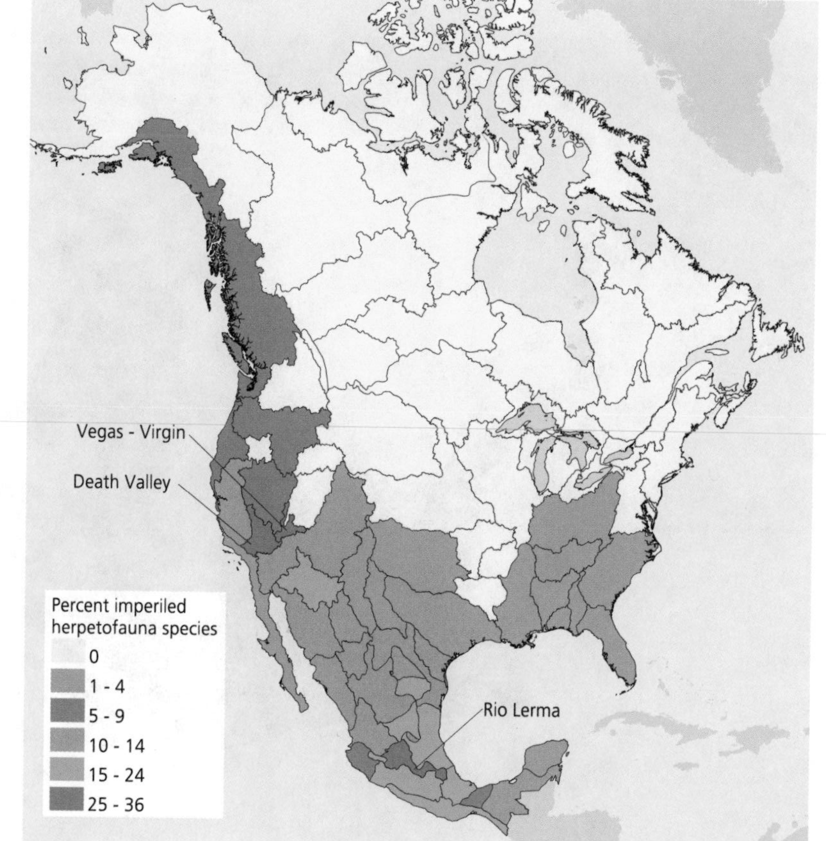

Source: Abell et al., 2000; The Nature Conservancy, 1997; Williams et al., 1989; CONABIO, 1998; Gonzales et al., 1995.
Projection: Lambert Conformal Conic
Central Meridian -96, Reference Latitude 40

Map 17
Number of Species Introductions into Inland Waters by Country

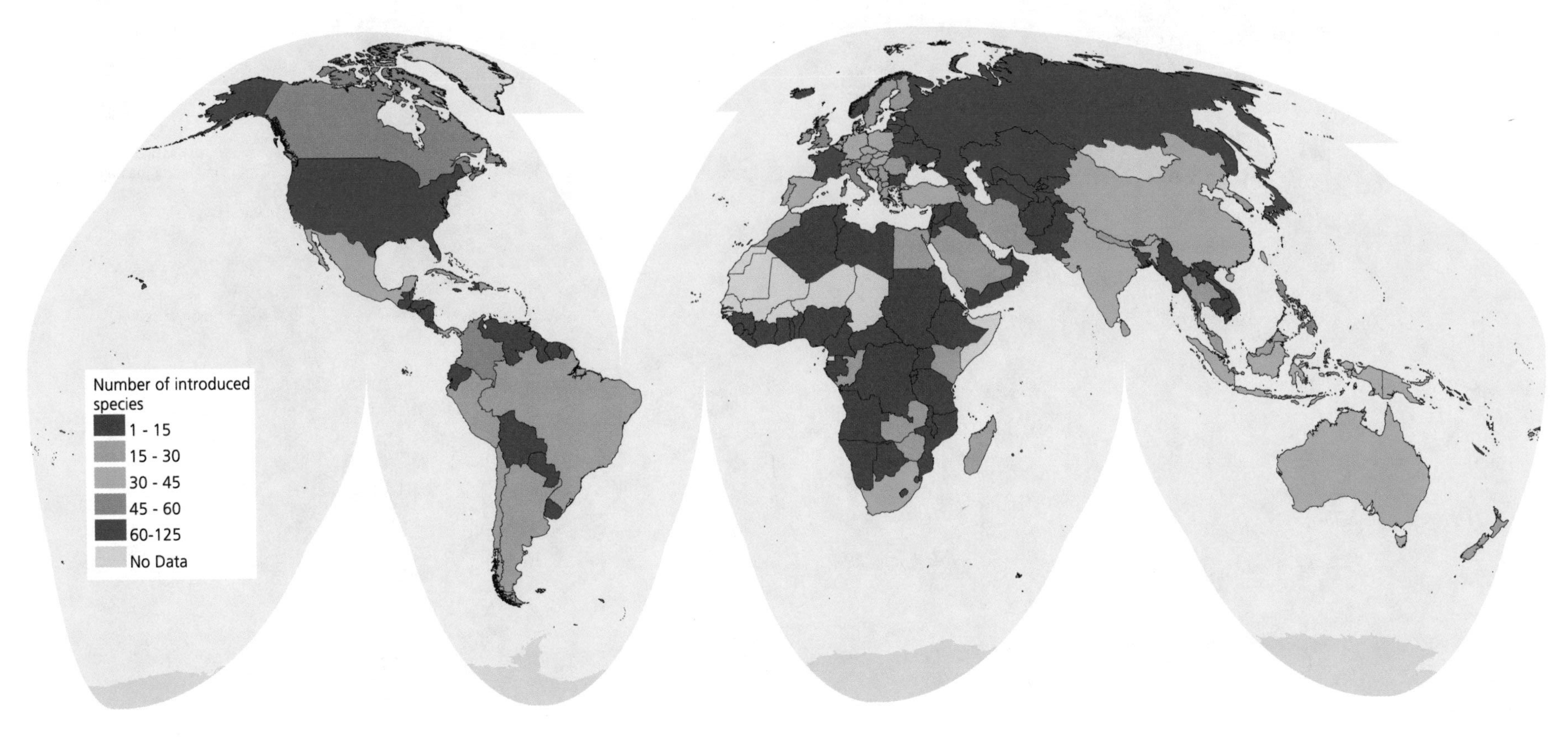

Source: DIAS, 2000; ESRI, 1996.

Projection: Interrupted Goode's Homolosine

Notes: Species introduced into large bodies of water adjoining more than one country (e.g. Black Sea, Caspian Sea, Great Lakes, Lake Kariba) are not included. Due to data limitations, values for the former Yugoslavia, Czechoslovakia and U.S.S.R are shown, rather than for the independent republics. Recent estimates of species introductions for the independent republics are: Armenia, 1; Croatia, 7; Estonia, 18; Georgia, 4; Kazakhstan, 5; Latvia, 1; Lithuania, 2; Russia, 13; Slovenia, 2; Turkmenistan, 1; Ukraine, 10; and Uzbekistan, 18. Azerbaijan, Belarus, Kyrgystan, Macedonia, Moldova, Montenegro, Serbia, Tajikistan, and Bosnia and Herzegovina have no records.

Map 18

Zebra Mussel Expansion and Water Hyacinth Presence in the United States

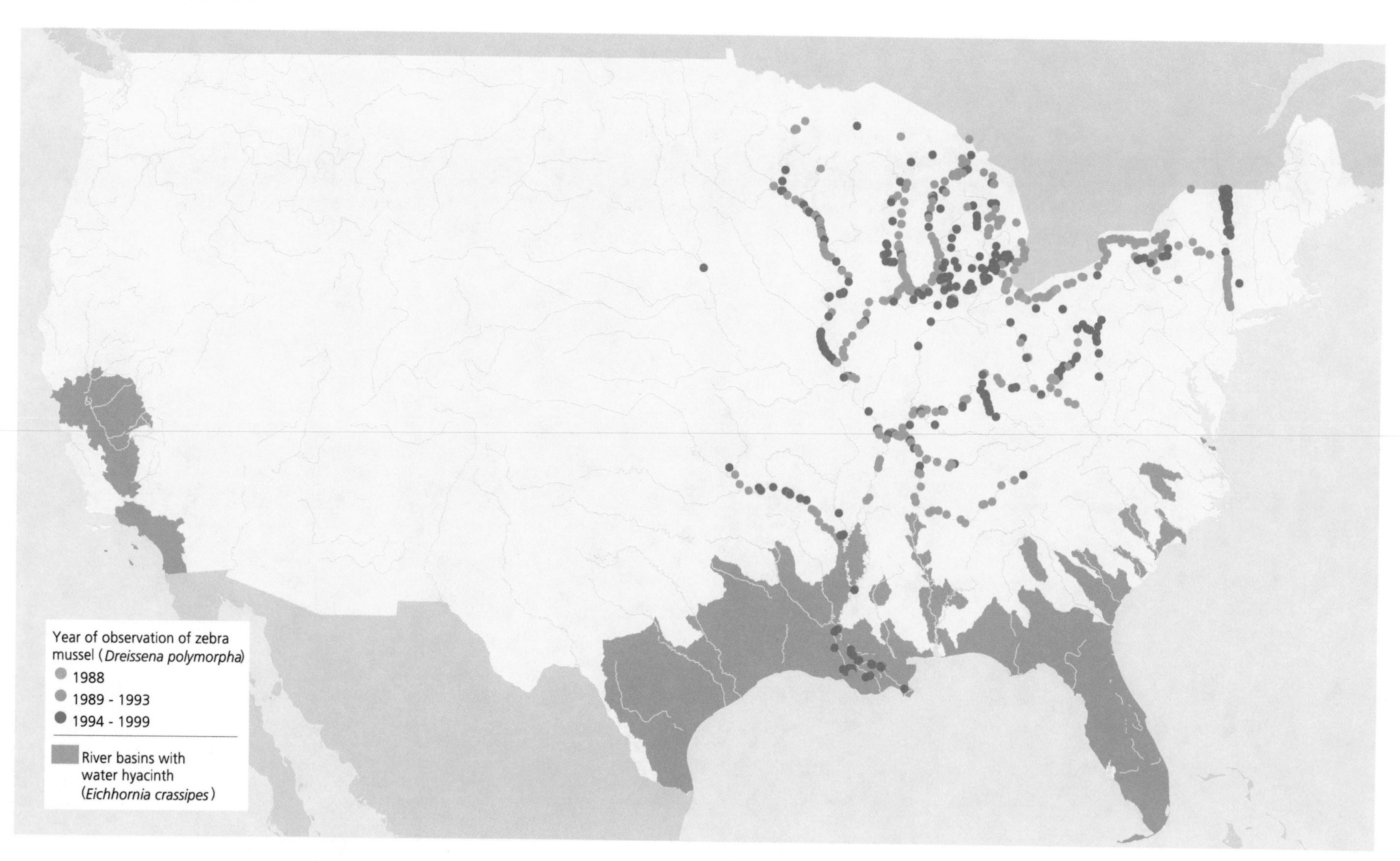

Source: USGS Zebra Mussel and Invasive Species Websites, 2000.

Projection: Interrupted Goode's Homolosine